Pest Control: Cultural and Environmental Aspects

AAAS Selected Symposia Series

Pest Control: Cultural and Environmental Aspects

Edited by
David Pimentel and John H. Perkins

Routledge
Taylor & Francis Group

LONDON AND NEW YORK

AAAS Selected Symposium **43**

First published 1980 by Westview Press

Published 2019 by Routledge
52 Vanderbilt Avenue, New York, NY 10017
2 Park Square, Milton Park, Abingdon, Oxon OX14 4RN

Routledge is an imprint of the Taylor & Francis Group, an informa business

Library of Congress Cataloging in Publication Data
Symposium on Environmental, Socioeconomic, and Political
   Aspects of Pest Management Systems, Houston, Tex., 1979.
   Pest Control.
   (AAAS selected symposium ; 43)
   "Papers ... presented at the symposium ... held at
the American Association for the Advancement of Science
meeting in Houston, Tex."
   Includes bibliographical references and index.
   1. Pest control--Social aspects--Congresses.
2. Pest control--Environmental aspects--Congresses.
I. Pimentel, David, 1925-   II. Perkins, John H.
III. Title.   IV. Series: American Association for the
Advancement of Science. AAAS selected symposium ; 43.
SB950.A2S95 1979   632'.9   79-18516

ISBN 13: 978-0-367-28278-3 (hbk)
ISBN 13: 978-0-367-29824-1 (pbk)

About the Book

The field of pest control research, of increasing importance in a world short of food, has been plagued for many years by a variety of problems, among them (1) the instability (including pesticide resistance) of many control techniques, (2) the continuing need for improved pest management methods to increase world food supplies, and (3) the environmental and social hazards of currently used pesticides. What historical or other factors affect the ability of science to generate useful new technologies to alleviate these three major problems? Are there barriers to cooperation among the different pest control specialists? This book attempts to answer these questions, examining past events and projecting likely impacts on contemporary pest management systems. The authors--sociologists, economists, lawyers, ecologists, political scientists, and pest control scientists--examine the social, economic, political, and ethical factors that are important in shaping pest management systems, as well as developmental patterns that show the importance of these factors in shaping today's systems.

About the Series

The *AAAS Selected Symposia Series* was begun in 1977 to
provide a means for more permanently recording and more
widely disseminating some of the valuable material which is
discussed at the AAAS Annual National Meetings. The volumes
in this *Series* are based on symposia held at the Meetings
which address topics of current and continuing significance,
both within and among the sciences, and in the areas in which
science and technology impact on public policy. The *Series*
format is designed to provide for rapid dissemination of
information, so the papers are not typeset but are reproduced
directly from the camera-copy submitted by the authors, without copy editing. The papers are organized and edited by
the symposium arrangers who then become the editors of the
various volumes. Most papers published in this *Series* are
original contributions which have not been previously published, although in some cases additional papers from other
sources have been added by an editor to provide a more comprehensive view of a particular topic. Symposia may be reports of new research or reviews of established work, particularly work of an interdisciplinary nature, since the AAAS
Annual Meetings typically embrace the full range of the
sciences and their societal implications.

WILLIAM D. CAREY
Executive Officer
American Association for
the Advancement of Science

Contents

List of Figures

List of Tables

About the Editors and Authors

David Pimentel *is professor of insect ecology and agricultural sciences at Cornell University and chairman of the Board of Science and Technology for International Development of the National Academy of Sciences. He is a member of the Advisory Panel (Genetics) of the Office of Technology Assessment and of the Energy Research Advisory Board, Department of Energy. His areas of interest include entomology, ecology, pest management, and agricultural science, and he has published numerous research papers in these fields plus five books.*

John H. Perkins, *associate professor of interdisciplinary studies at Miami University, is currently an honorary research associate at the Division of Biological Control, University of California-Berkeley, and is working on a book on the history of entomology. He is a member of the executive committee of the American Society for Environmental History and of the Technology Studies and Education Committee, Society for the History of Technology. He served as principal staff officer for the Study on Problems of Pest Control (1971-1974) and as member and head analyst of the Subpanel on Plant Protection, Study Team on Crop Productivity, World Food and Nutrition Study (1976-1977) of the National Academy of Sciences. Among his many publications is* Contemporary Pest Control Practices and Prospects, *Vol. I (National Academy of Sciences, 1975), of which he is coauthor.*

David Andow *specialized in insect ecology and population genetics at Brown University. He is presently an NSF Graduate Fellow in Ecology at Cornell University and is a member of the Society for the Study of Evolution.*

Rada Dyson-Hudson *is associate professor of anthropology at Cornell University. Her areas of specialization are ecology, evolutionary biology and human ecology. She is a Fellow of the American Anthropological Association and a John Simon Guggenheim Foundation Fellow. Among her publications is "Food Production System of the Karimojong of East Africa," in* African Food Production Systems *(P. McLoughlin, ed.; Johns Hopkins Press, 1970). She has also contributed to* Scientific American *and* The Journal of Asian and African Studies.

David Gallahan *is doing postgraduate work in ecology and systematics at Cornell University, specializing in evolutionary ecology. He has written articles on pesticides and cosmetic standards in foods, pest control strategies, and benefits and costs of pesticide use.*

Judith Hough *is a graduate student in entomology at Cornell University. Her area of specialization is insect-plant interactions, and she served as staff officer for the National Academy of Sciences study* Pest Control: An Assessment of Present and Alternative Technologies. *Her other publications include papers on pest control strategies and cost/benefit analysis of pesticide use. She is a member of the Entomological Society of America.*

Mary Ann Irish *received her degree in agriculture and life sciences from Cornell University, and her specific area of specialization is pest management.*

Stuart Neil Jacobson *is a graduate student in agronomy at Cornell University. He has done work there on land treatment of waste water and on herbicide biodegradation. His area of specialization is soil microbiology, and he is a member of the American Society for Microbiology.*

Susan F. Kroop *received her degree in biology from Cornell University and is now a medical student there.*

John Krummel *is a research associate in entomology at Cornell University where he has worked and published on the topics of population and agricultural ecology and pesticide use. He is a member of the Ecological Society of America and the Society for the Study of Evolution.*

Anne M. Moss *is a medical student at George Washington University, specializing in neurobiology and behavior.*

Ilse Schreiner *is a graduate student in entomology at Cornell University where she is doing work in insect ecology.*

Michael D. Shepard, *whose academic background is in natural resources conservation, has written articles dealing with organic agriculture and renewable energy technologies and is primarily interested in the development of these fields in relation to solar energy.*

Jerry D. Stockdale, *associate professor of sociology at the University of Northern Iowa, has published many papers dealing with social change, rural development, and agricultural technology and its social and environmental impacts. He contributed as a study-team member to the Environmental Studies Board of the National Academy of Sciences, dealing with the problems of pest control, particularly for corn and soybeans. A member of the American Sociological Association and the Rural Sociological Society, he co-authored, with Gould Colman,* Area Development through Agricultural Innovations *(West Virginia University Press, 1977).*

Todd E. Thompson *is a graduate student and research assistant in agronomy at Cornell University. A member of the American Society of Agronomy and the Crop Science Society of America, he has published a study dealing with computer simulation of alfalfa growth.*

Billy G. Vinzant *is a systems engineer at the General Electric Company working on controlled environment agriculture. He most recently published "The Effect of IR Radiation in Growth Chambers" (General Electric Co., 1978).*

A. Dan Tarlock, *visiting professor of law, University of Chicago (1979) and professor of law, Indiana University, specializes in natural resources law. He was a member of the Cotton Study Team of the National Academy of Sciences (1973-1975), and has published two books in this field:* Water Resource Management *(with Charles J. Meyers; Foundation Press, 1979) and* Environmental Law and Policy *(with John and Eva Hanks; West Publishing Co., 1974).*

Preface

Papers included in this book were presented at the Symposium on "Environmental, Socioeconomic, and Political Aspects of Pest Management Systems" held at the American Association for the Advancement of Science meeting in Houston, Texas, January 1979. The objective of the Symposium was to examine the social, economic, political, and environmental factors that are important in shaping pest management systems. At this Symposium for the first time sociologists, economists, lawyers, ecologists, science policy analysts, and pest control scientists exchanged views concerning the variety of problems that plague pest control. This knowledge may help society stabilize pesticide-based control techniques, improve pest management methods to increase food production in the United States and world, and reduce pesticide hazards to public health and the environment.

We appreciate the assistance of Ms. Nancy Goodman and Mr. Michael Burgess in assembling and indexing and Ms. Beth French in typing some parts of the book. We also thank Dr. Kathyrn Wolff and Ms. Joellen M. Fritsche of AAAS for their help in seeing that the results of the Symposium were published.

Pest Control: Cultural and Environmental Aspects

John H. Perkins, David Pimentel

1. Society and Pest Control

Abstract

Policy making for pest control is set in a complex of
biological, environmental, and cultural concerns. Total
losses to pests of food supplies are large, about 45%. Pest
population sizes are heavily influenced by agricultural
practices. Control of pests in agriculture can play a role
in alleviating world hunger, but important sociopolitical
factors must also be recognized. The framework for
analyzing pest-control activities must include detailed
examinations of the community of pest-control experts, the
users of the expertise, and the rest of society.

Introduction

The objective of this book is to examine the relationships between pest control research and practices on the one
hand and cultural factors on the other. By cultural factors
we mean the economic, social, political, environmental, and
philosophical aspects of American society. Often, scientists
and others forget that all technologies are forged in and
shaped by a complex social setting. It is necessary to
understand the relationships among the various cultural
factors that influence and determine pest control research
and practices if we are to establish adequate public
policies for this problematic set of technologies.

Efforts to create and implement better pest control
practices are currently set in a background of concerns about
the global environment and the adequacy of world food distribution and production. In this chapter, we examine these
background issues. They include losses to pests, the ecological causes of pest outbreaks, human population growth and
food demand, the environmental resources needed for world

food production, and the cultural context of pest-control practices.

World Food Losses to Pests

Food losses to pests worldwide are large. It is esti-mated that 35% of potential production is lost to pests (Cramer, 1967). This loss is occurring in spite of pesti-cidal and other control programs. The primary pests are insects, diseases, and weeds. However, under certain cir-cumstances, particularly in the tropics and subtropics, mammal and bird losses may be important. But these losses are still low compared to the three major pest groups.

Postharvest food losses to pests range from about 9% in the United States (USDA, 1965), to 10-20% in other parts of the world (NAS, 1978). The prime pests of harvested foods are microorganisms, insects, and rodents.

Adding postharvest food losses to preharvest food losses, the total world food losses due to pests are esti-mated to be about 45%. Thus, the pest populations are con-suming and are destroying nearly one-half of the world food supply. Surely this is a loss that we cannot afford as we face world food shortages and an ever-increasing world popu-lation. A recent study by the National Academy of Sciences (1977, p. 102) estimated that if 20% of the current pre-harvest losses of rice could be saved, the additional 56 million tons of rice available could provide adequate calories for 177 million people per year. The rice saved would almost be sufficient to feed the combined populations of Japan and Bangladesh.

Ecological Basis for Pest Outbreaks

The ecological basis for insect pest, pathogen, and weed problems is complex. Pest outbreaks are often the result of a combination of ecological factors. One of these cases is the monoculture of crops. Natural ecosystems tend to evolve toward stable climax communities for each particu-lar habitat in the world. For agricultural production, how-ever, the natural plant community is removed and destroyed, and is replaced by a single crop species. As soon as the land is cleared of the natural vegetation, man's battle with what he terms pests begins. The seeds that are planted germinate, but so do hundreds of seeds of other plant species that lay in the soil, some of which may have remained dormant for many years. In addition, various microorganisms are present in the soil or may drift in the wind. Insects may be present or fly in from other locations, with all of these insects tending to attack the crop.

One aspect of the monoculture problem is that the larger the area that is planted to a single crop, the greater the potential for pest problems. This also relates to the problem of continuous culture or a monoculture of one crop in one location for several years. When crops are maintained in the same area year after year, pests associated with the crop tend to increase in number and in severity. This is true of cole crops. For example, if they are cultured for several years in the same soil, club root (<u>Plasmodiophora</u> <u>brassicae</u>) organisms increase rapidly and can totally ruin production (Walker et al., 1958). The same is true of the corn rootworm that is now the primary pest of corn. When corn is planted following soybeans or small grains, corn rootworms are not a problem. Crop rotations, however, can sometimes increase a problem with black cutworms and wireworms. The use of crop rotations is not, therefore, a complete answer to insect problems in corn (NAS, 1975, pp. 53-55).

Some pest problems occur when crops are introduced into new biotic communities. For example, when the potato, which originated in Bolivia and Peru (Hawkes, 1944) was introduced into the southwestern United States, it acquired a serious pest, the Colorado potato beetle. Native to the United States, this beetle had originally coevolved with and fed on wild sand bur (Elton, 1958). When the potato was introduced into the southwest, the beetle spread onto it. Because the potato had never been exposed to the beetle, it lacked any natural resistance to it. Since then, this insect has become the most serious pest to potato in the world and has accompanied the plant as its cultivation spread to other areas.

In addition to introducing crops into new locations and having pest problems develop, the introduction of pest species is one of the most important causes of pest problems. A few examples of newly introduced species that became pests are the European rabbit that was introduced into Australia, the Japanese beetle that was introduced into the United States, the gypsy moth, Dutch elm disease, and water hyacinth.

One of the most important ecological factors involved in pest problems is the breeding of susceptible crop genotypes (Lupton, 1977). An example of this is sorghum. On a susceptible strain of commercial sorghum, the mean rate of oviposition (eggs per generation) of the chinchbug was about 100. On a resistant strain, however, the mean oviposition was less than 1 (Dahms, 1948). In this instance, the animal feeding was reduced by 99% on the resistant plants. This has had dramatic effects on the population dynamics of chinchbugs.

Much has been written about diversity and its influence on the stability of pest populations. Frequently, outbreaks of insect pests in agriculture have been attributed to crop monocultures. For example, Marchal (1908) wrote that when man plants a vast extent of the country with certain crops, while excluding others, he offers to the insects feeding on these plants favorable conditions for their explosive increase. This has been documented with experimental studies on the plant <u>Brassica</u> oleracea by Pimentel (1961a), Tahavanainen and Root (1972), Cromartie (1975), and Root (1973 and 1975).

Based on numerous examples, it is clear that many parasites have the genetic variability to evolve and overcome what might be termed single factor resistance in their host. For example, parasitic stem rust and crown rust have been found to overcome genetic resistance bred into their oat hosts. Since 1940, oat varieties have been changed in the Corn Belt region every five years to counter the changes in the races of stem rust and crown rust (Stevens and Scott, 1950; van der Plank, 1968). In experiments with an animal and simulated plant model, genetic stability has been demonstrated in an animal-plant relationship when six resistant factors (genes) were present in the plant population (Pimentel and Bellotti, 1976).

Another factor contributing to pest outbreaks is plant spacing. In cultivated fields, crop plant densities are carefully controlled to obtain the maximum population possible for optimal growth, resulting in maximum economic yield. Seldom are the spacings of such plants similar to those in the wild. The new plant spacings often result in an ecological situation that encourages pest outbreaks (Pimentel, 1961b).

All plant feeding insects have specific nutrient requirements. Altering the nutrient level in the soil and then in the host plant influences the pests that are feeding on the plant. An improvement in nutrients often results in the parasite increasing, and a decline in the nutrients results in the reverse. For example, Haseman (1946) reported that the grain aphid, feeding on small grain plants with high nitrogen produced an average of 33 progeny per aphid, whereas on plants with a low level of nitrogen progeny production averaged only 13 per aphid.

Another factor contributing to pest outbreaks is the particular host plant association that is employed by growers. Some pests, for example, can attack and feed on several species of host plants. These pests can move from one host plant to another when one of the host plant populations declines in abundance for some reason. For example, plant bugs feed on alfalfa and cotton and when alfalfa is mowed for hay and eliminated as a food source, the bugs will move onto cotton in large numbers (Stern, 1969).

Some pesticides may alter the physiology of crop plants and therefore make them more susceptible to pest attack. For example, herbicides have been found to increase insect pest and pathogen problems associated with corn (Oka and Pimentel, 1976).

Human Population Growth and Food Demand

The ecology of pest outbreaks is related to the ecology of the human population. At no time in history have humans so dominated the environment of earth; yet this is a very recent phenomenon. For more than 99% of all the time that man has been on earth, the maximum population reached was an estimated 200,000, or only one one-thousandth the population of the United States today. At current reproductive rates, the world population is growing by an additional 200,000 people per day.

Annual population growth during most of human history was less than .01% (NAS, 1978), considerably less than its current level. This is the prime reason why the world population remained so small for such a long period of time. Man survived as a hunter-gatherer throughout most of human existence. The hunter-gatherers were, in a sense, astute ecologists who recognized the limitations of the environment in supporting the human population. They controlled their numbers and adjusted them to the resources that were available for their support. Interestingly, today the arguments for population control range over abortion, economics, political structure, and so forth; but our ancestors, with simple political systems and no economic development, were able to achieve effective control of their numbers. Apparently, when groups of people recognize the need for population control, they can achieve this control without education, economic development, and complex political systems.

The first major increase in human numbers occurred with the discovery of agriculture, the domestication of crops and animals. When agriculture became established about 10,000

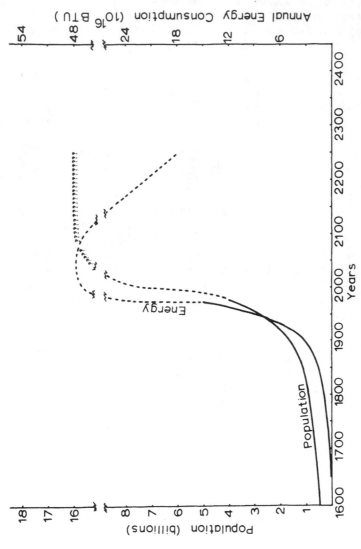

Figure 1. Estimated world population numbers (———) from 1600 to 1975 and projected numbers (– – –) (????) to the year 2250. Estimated fossil fuel consumption (———) from 1650 to 1975 and projected (- - - -) to the year 2250. Reprinted
with permission from Pimentel et al., "Energy and Land," in Science, Vol. 190,
21 Nov. 1975, pp. 754-761. Copyright 1975 by the American Association for the
Advancement of Science.

years ago, the supply of food increased and at the same time the quantity of food stabilized. This contributed to and allowed an increased rate of growth for the world population.

The dramatic increase in world population occurred after 1700 (Figure 1). The rapid population growth coincided with the exponential use of fuels. Fossil energy has been used for disease control operations, and to improve agricultural production in order to feed the growing population. Both the effective control of human diseases and the increased food supply have contributed significantly to the rapid growth in human numbers (NAS, 1971). Of these two factors, the evidence suggests that the reduction in death rates through effective public health programs is the prime cause (Freedman and Berelson, 1974). The effective control of malaria-carrying mosquitoes by DDT and other insecticides is a good example (Note, substantial quantities of energy are required for the production and application of pesticides).

After spraying with DDT in Ceylon in 1946 and 1947, the death rate fell in one year from 20 to 14 per 1000 (PEP, 1955). A similar dramatic reduction in death rates occurred after DDT was used in Mauritius, where death rates fell from 27 to 15 per 1000, and population growth rates increased from about 5 to 35 per 1000 (Figure 2).

The picture relative to malaria is now changing in the world. From 1961-63, the lowest incidence of malaria occurred in many parts of the world. For example, in India during that period, there were about 100,000 cases of malaria. In 1978, there were 50 million cases of malaria reported in India (Figure 3). The reason for the explosive increase of malaria during the last few years has been an increased level of pesticide resistance that has evolved in the mosquito vector population. The increased rate of resistance that has been developing in the mosquito population is due to the increased use of insecticides in agriculture to help increase food production. Thus, the attempt to provide more food to reduce world food shortages has, in fact, resulted in deaths from another cause, and that is disease (malaria).

Most of the application of insecticides in agriculture is on crops. These food crops provide most of the nutrients consumed by the human population. About half of these food nutrients come from cereal grains, including wheat, rice, corn, millet, sorghum, rye, and barley. About 90% of all the food for humans comes from these cereal grains plus cassava, sweet potato, potato, coconut, banana, common bean, soybean, and peanut (NAS, 1978).

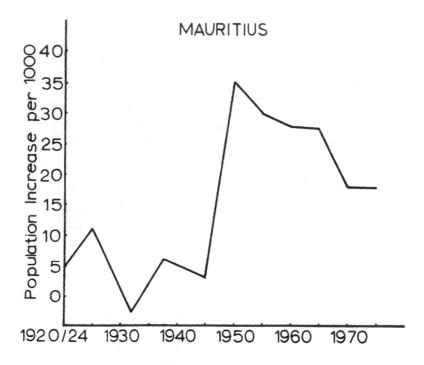

Figure 2. Population growth rate on Mauritius from 1920 to 1970. Note from 1920 to 1945 the growth rate was about 5 per thousand whereas after malaria control in 1945 the growth rate exploded to about 35 per thousand and has since very slowly declined. (Reprinted with permission from Pimentel & Pimentel, Food, Energy and Society, Edward Arnold, Publishers, 1979.)

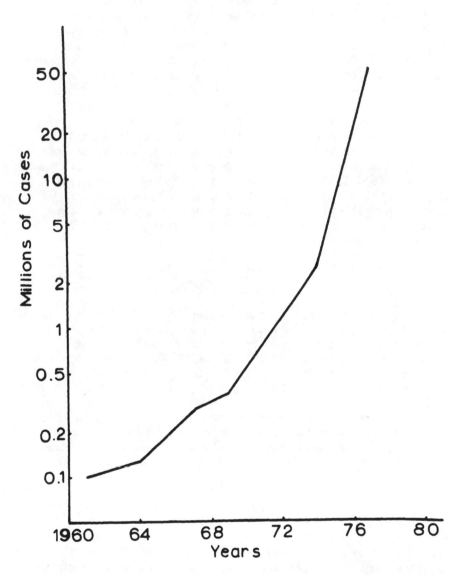

<u>Figure 3.</u> The estimated number (log) of cases of malaria
reported in India from 1961 to 1977 (Harrison, 1978).

In many parts of the world there is insufficient production of food to support healthy human life. In fact, it
is estimated that about a half-billion humans are mal-
nourished today (FAO, 1974). Thus, at the current production
levels, not all humans receive sufficient food. Clearly,
the loss of nearly one half of potential production to pests
is a world problem that demands attention. It must be
emphasized, however, that the biological problems of pest
losses are intimately intertwined with a complex of cultural
factors. Decreasing losses to pests would not automatically
alleviate malnutrition (NAS, 1977). Likewise it is not
necessary to reduce pest losses in order to eliminate some of
the worst malnutrition today; changes of the social structure
in which food is produced and distributed could accomplish
the same thing (Lappe and Collins, 1977). Our argument is
simply that alleviation of the biological losses to pests
could be a significant factor in improving levels of global
nutritional well-being.

Environmental Resources
Utilized in Food Production

To gain some idea of the challenges of feeding a rapidly
growing world population, estimates are made of the animal
and vegetable matter production relative to land, water, and
fossil energy constraints. This analysis assumes a present
population of 4 billion, 6 billion in the year 2000, and 16
billion for the year 2100.

Many peoples of the world desire to eat and live as the
people of the United States. Hence, in the first analysis,
we calculate land and energy required to feed a population of
4 billion a U.S. high protein-calorie diet produced with U.S.
agricultural technology.

In the United States about 160 million hectares are
planted to crops (USDA, 1977). With a U.S. population of
about 215 million, this averages about 0.7 hectare planted to
crops per capita. Since about 20% of our crop yield is
exported, the estimated arable land per person is about 0.62
hectare (USDA, 1977). The world arable land resources are
about 1.5 billion hectares (FAO, 1973). With 4 billion
humans in the world today, the per capita land available is
only 0.38 hectare. In the United States, 0.62 hectare of
land plus a high energy agricultural technology are necessary
to produce the high protein-calorie diet that is consumed.
Hence, in the world today, there is insufficient arable land
(even assuming that energy resources and other technology are
also available) to feed the current world population a diet

similar to and produced in the same manner as that consumed in the United States.

In this analysis of land resources, fossil energy was assumed to be adequate. Unfortunately, fossil energy is in limited supply for food production. This can be put into perspective with the following analysis. If petroleum were the only source of energy for food production, and if all petroleum reserves were used solely to feed the world population, the 66,000 billion liter oil reserve in the world would last a mere 13 years (Pimentel et al., 1975). Both the land and energy estimates indicate that the human population has already reached a density too great for the arable land and energy resources that are required to feed the world population a U.S. diet utilizing U.S. technology. Both estimates were made from known arable land and petroleum resources. If we include potential arable land and potential petroleum resources, the situation appears to be improved. It should be pointed out that a population of only 4 billion was used in the calculations. The world population is now 4.3 billion and is projected to reach more than 6 billion in the next 21 years.

We will make another analysis recognizing the constraints of land and energy resources while food demand increases with a rapidly growing world population. The focus is on both the animal and vegetable foods and their availability to the human population. Total animal protein consumption by man today amounts to about 25% of the total protein supply consumed by the world population. Cereals contribute nearly half of the total protein supply consumed by man. There would be sufficient foods for all people throughout the world if pests and other production factors were under control. Even with pest losses and other types of loss, there should be adequate amounts of food available if it were equitably distributed to all peoples.

Livestock production may be increased 30% by the year 2000 through reduced overgrazing and the use of better pasture plant species and application of limited amounts of fertilizer under certain advantageous conditions. But to hold the per capita food supply in the year 2000 at 1975 levels will require a 66% increase in legumes, a 100% increase in other vegetables, and a 75% increase in cereals (Pimentel et al., 1975). This 75% increase in the next 25 years is technically feasible; the 66% increase for legumes and the 100% increase for vegetables appears less likely.

To feed the 16 billion humans predicted for the year 2100, utilizing a similar diet to that consumed by the world population in 1975, will require significant increases in food production. For example, legumes must be increased by 173%, vegetables 233%, and cereals 330% (Pimentel et al., 1975). With the resources of land, energy, and water that are available, these increases appear to be doubtful.

One means of increasing the total amount of food available to man would be to reduce the amount of vegetable and other animal products that are currently fed to livestock. An estimated 51 million metric tonnes of protein suitable for man's use were fed to the world's livestock in 1975 (Pimentel et al., 1975). This 51 million metric tonnes that is fed livestock is nearly equal to the total cereal protein available to man for 1975. Therefore, if man could switch from consuming quite as large an amount of animal products to consuming more plant products there would be more food available for the world's population.

With careful management of land, water, energy, and human resources, and cooperation among nations of the world, we believe that it is possible to maintain current per capita levels of food supplies for the next 21 years, as the world population increases to more than 6 billion humans. Serious malnutrition already exists with some half billion humans. Efforts are needed to eliminate this deficiency by better food production and distribution.

Science and technology in pest control can help man to overcome future food crises that face him as his numbers rapidly increase, but the necessary solution to the well-being of mankind will likely also require a more equitable distribution of resources and effective population control. The problems associated with pest control are inextricably bound, therefore, to complex biological and sociopolitical aspects of human life. In the next section, we turn to a framework for analyzing pest-control activities in their complex, cultural context.

The Cultural Context of Pest-Control Practices

Pest Control is a Cultural Activity

Pests are living organisms that, when present in sufficient numbers, cause events and processes that people dislike. Pests are not limited to any taxonomic class and include weeds, animal and plant pathogens, insects, vertebrates, and others. People dislike pests because they cause damage and discomfort to themselves and their possessions.

Biologically, little distinguishes a pest from a non-pest; instead, the most important factors dividing pests from other species are based on human judgments and preferences. Evaluations of pest-control practices must therefore be based on an understanding of both the biological and cultural attributes of those organisms we label as "pests."

Biologically, pest control must be based on an ecological understanding of the factors affecting pest distribution and abundance. Ecological analysis indicates that organisms become sufficiently abundant to arouse human ire because conditions are suitable for their reproduction and survival. Many of these conditions are themselves the results of human activities such as agriculture. We also understand pests ecologically as competitors: for example, insects compete directly with us by feeding upon our crop plants and livestock; weeds compete indirectly by crowding our crop plants for light, nourishment, and water.

Culturally, pest control must accommodate a variety of economic, social, political, philosophical, aesthetic, and ethical factors important to society. A multitude of cultural factors join to create constellations of concerns surrounding and in part defining each specific pest problem. Pest control practices must be compatible with cultural factors originating from economic, political, intellectual, and social activities. They must also accommodate values and assumptions about the roles of humans in the natural and social orders. Indeed, pest control is based upon values and traditions fundamental to our culture. Dedication to private property and individual freedom, for example, are deeply enmeshed in the concepts of what constitutes "good" pest-control technology in the United States.

One of the major problems with efforts to move away from heavy reliance on pesticides has been the mistaken perception that the issues involved were largely technical. If technical matters alone were at issue, then more money to pest-control scientists for innovation and to extension personnel for education would undoubtedly suffice. The difficulties of implementing non-chemical control suggest that the problem is not simply one of fostering invention-diffusion-adoption. The quest to induce change in pest-control practices requires a deeper appreciation of its cultural foundations. It is possible that changes in pest-control practices will come only after adjustments are made in our culture. The magnitude and quality of such changes are largely unexplored, but it is likely they will be neither trivial nor easy.

The papers presented in this Symposium are a continuation of efforts begun largely within the past ten years to understand how the pest control enterprise functions within its cultural context. Human cultures are so complex that it is necessary to simplify them in order to see fundamental patterns. The simplification scheme inherent in this volume is shown in Figure 4. Culture is embedded within the material world (the "environment"). Three elements of society are functionally distinguishable in pest control activities: (1) experts who develop new techniques; (2) users of pest control techniques, especially farmers; and (3) the rest of society, those who neither invent techniques nor use them directly in their daily activities. Each of these social groups has important interactions with the other two; they are formally tied together with traditions and laws (Government and Policy).

Pest-Control Experts and Their Expertise

Pest control in its earliest forms was not an activity involving special expertise. Prehistorical hunter-gatherer cultures and the first agricultural communities undoubtedly practiced the destruction of weeds, some insects, and other pests by simple mechanical and physical means. More complex means of pest control were known in early historical times (Smith et al., 1973). The late nineteenth century was a period of transformation of pest control from an art known to almost everyone to a science developed and implemented by a group with special knowledge (expertise). During the twentieth century, pest control scientists developed into a recognizable community distinguished by their education, places of employment, and daily work patterns (Howard, 1930). These professional students of pest control are now commonly regarded as the people responsible for guiding their fellow citizens through the intricacies of controlling unwanted organisms with efficiency and safety. In return, the professionals expect a certain modicum of honor, recognition, and privilege (monetary and otherwise) for their efforts. The relationships between the professional pest-control scientists and the rest of society are thus typical of those associated with other groups possessing expert knowledge.

Despite the importance of pest-control expertise in developing nonchemical control practices, little has been done to learn how the community of experts functions. We know how they get their formal education, but only the rudiments about the nature and origin of their attitudes. We know where they work, but we know little of how their employing institutions affect their products. We know they relate to non-experts of many types, but we know little of how these

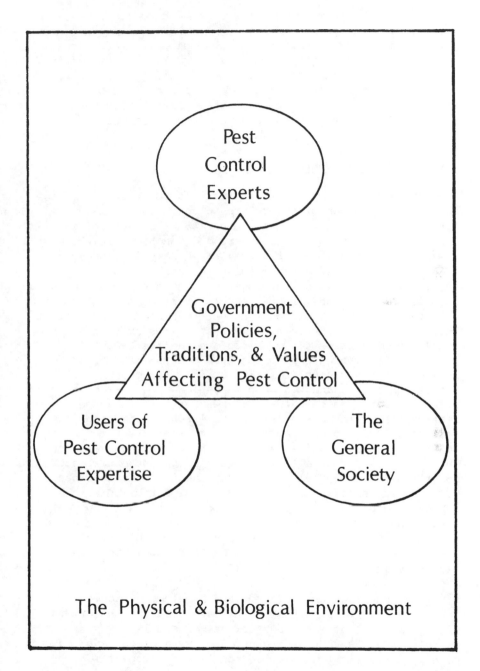

Figure 4. Pest control in its cultural and environmental contexts.

relationships affect their research and advocacy of pest-control technologies. Finally, we know they relate to each other through a wide variety of professional interactions, but we know little of the dynamics of such interactions and their impacts on the content of the expertise. Unless we develop a better understanding of the functioning of this important community of experts we will have little basis for anticipating the types and rates of innovation they are likely to produce.

The Users of Pest-Control Expertise

Pest-control expertise is developed to be used as a tool in the exploitation of natural resources. The division of labor in complex societies has separated the users of exper-tise from the developers. Farmers and ranchers are the most easily recognized group of users, especially in the indus-trialized world. In the United States, the origin of the institutions to develop expertise in pest control accompanied the rise in the need for that expertise by commercial farmers and ranchers. The research arms of the USDA, the land-grant universities, and the chemical industry were created and are still the home bases for most professional pest-control scientists. Other groups using pest-control expertise include foresters, public health officials, non-agricultural industrial firms, and home owners/gardeners. The importance of controlling insect-borne diseases for the public health has resulted in some specialized research institutions in medical schools to develop the needed expertise for public health officials. Aside from this exception, however, the general rule is that all people wishing to use expert pest control must usually obtain it from institutions developed to serve the special needs of commercial agriculture. The farming sector thus occupies an important place in the cul-tural context of pest-control activities. Most pest-control expertise, furthermore, is generated in the industrialized world. Severe problems can attend its transfer to less industrialized countries (NAS, 1977).

Our knowledge of the needs and problems of commercial agriculture is considerably higher than that of the community of pest-control experts, but deficiencies still exist. For example, we know that pest control plays an important economic role by increasing quantity and quality of crop yields, by providing insurance against catastrophic losses, and by freeing farmers from restraints in adopting other types of technology such as irrigation, machinery, or fertilizers. We have less detailed knowledge about the role pest control plays in the overall activities of a farmer with a highly diversified operation including different crops, livestock,

and "sideline" businesses such as warehouses, gins, and so
forth. We know considerably less about how pest-control
expertise plays differential roles in farmers of varying
economic, social, racial, and cultural backgrounds. Finally,
the agricultural community is still undergoing a transformation begun over a century ago in which capital is substituted
for labor, the number of people in agriculture decreases, and
the average size of farm operation increases. This transformation has already profoundly affected farmers' needs for
expertise in the industrialized world, but we have little
understanding of what the future will bring.

The Rest of Society

Those people who are neither professional pest-control
scientists nor major users of their expertise constitute a
highly heterogeneous group recognizable by what its members
don't do instead of what they do. If the size of this group
was small compared to the other two, then it might not be
particularly important. It isn't: about 95% of the individuals in the United States are in it, and it therefore is
a highly significant political and economic force. The members of the first two groups can ignore the interests of the
overwhelming majority of their fellow citizens only at considerable peril to their own future freedom of action. In
less industrialized countries, the agricultural sector is
still frequently the largest social group. Even so, the
general society may exert considerable pressure on the shape
of acceptable technology.

We know something about the general society's interests
in pest control, but that knowledge provides at best a conflicting guide to policy needs. For example, the general
society as consumers wants low prices for food and fiber and
some protection from deadly insect-borne diseases and other
"nuisance" pests. At the same time, the general society has
an interest in keeping the environment from being contaminated with pesticide residues. Pesticides have been heavily
used as a means of assuring the first set of goals, but their
uses interfere with the second objective. Furthermore, it is
unclear that improved pest control is the most efficient
method by which to help lower the prices of food and fiber.
The set of issues surrounding food prices, environmental
quality, and pest control is just one of many affecting the
general society, but it illustrates the point that the
interests of the general society are complicated and perhaps
inherently filled with contradictions. Our understanding of
the nature, magnitude, and significance of these contradictions is poor.

The Problems Addressed
in This Symposium

This small volume will not provide answers for all of
the questions raised in this introduction. Indeed the
reader may come away somewhat frustrated because the papers
tantalize more than answer these difficult problems. The
editors would prefer to view this work as the outline of an
analytical framework which needs a great deal more work before we can gain a better view of how pest control fits
within culture and what might be done to alleviate the
problems plaguing the past three decades. The analysis here
is confined largely to the United States. The editors regret
it was not possible to adopt a more global perspective, but
constraints of time and facilities made the restrictions
necessary.

The volume begins with an examination of the community
of experts concerned with insect control (Chapter 2 by
Perkins). The method is historical; and the focus is on
the efforts to innovate in agricultural entomology, the pest-
control science subjected to the most intense controversy
during recent years. He demonstrates that the efforts to
innovate in entomology have been closely tied to a variety
of cultural factors. Chapter 3 (Headley) provides a historical overview of how the economic milieu of agricultural
production in the U.S. has changed since 1945. The high
suitability of pesticides for the economic structure of our
farming industries dramatizes why it has been so difficult
to move away from their heavy use. Pimentel et al. (Chapter
4) provide a comprehensive yet conservative overview of the
aggregated indirect costs of pesticides to the farming community and the general society. Their arguments and figures
demonstrate on economic grounds that the problems associated
with pesticides are by no means trivial and insignificant.

It is generally recognized that pesticides must provide
benefits to justify the risks and problems associated with
them. Krummel and Hough (Chapter 5) probe the assessment of
risks and benefits with the combined methods of cost
accounting and environmental impact assessment. They demonstrate that the assessment of costs and benefits of pest
control activities is complex and that what may first appear
to be highly favorable benefit/cost ratios are diminished by
more holistic analysis. Stockdale (Chapter 6) presents an
appraisal of the concepts needed to guide social and political
policy affecting pest control activities. He argues that we
need to know where we want to go before effective policy can
be established. Tarlock (Chapter 7) concludes the volume with
an examination of the major law governing pest control

activities. He reaches the important conclusion that the law as presently constituted is not well suited to promoting the adoption of pest-control techniques less dependent on chemicals.

References

Cramer, H. H. 1967. Plant protection and world crop production. Pflanzenschutznachrichten 20(1): 1-524.

Cromartie, W. J., Jr. 1975. The effect of stand size and vegetational background in the colonization of cruciferous plants by herbivorous insects. J. Appl. Ecol. 12: 517-533.

Dahms, R. G. 1948. Effect of different varieties and ages of sorghum on the biology of the chinch bug. J. Agr. Res. 76(12): 271-288.

Elton, C. S. 1958. The Ecology of Invasions by Animals and Plants. Methuen, London. 159 pp.

FAO. 1973. Production Yearbook 1972. vol. 26. Food and Agriculture Organization, Rome.

FAO. 1974. Assessment of the World Food Situation. World Food Conference, Food and Agriculture Organization, Rome.

Freedman, R., and B. Berelson. 1974. The human population. Sci. Am. 231(3): 30-39.

Harrison, G. 1978. Mosquitoes, Malaria and Man: A History of the Hostilities Since 1880. E. P. Dutton, New York. 314 pp.

Haseman, L. 1946. Influence of soil minerals on insects. J. Econ. Entomol. 39: 8-11.

Hawkes, J. G. 1944. Potato collecting expeditions in Mexico and South America. Imp. Bur. Plant Breeding Genetics 633.491-1.524(8), School of Agriculture, Cambridge, England.

Howard, L. O. 1930. A History of Applied Entomology (Somewhat Anecdotal). The Smithsonian Institution, Washington, D. C. 564 pp.

Lappe, F. M. and J. Collins. 1977. Food First. Houghton Mifflin Co., Boston. 466 pp.

Lupton, F.G.H. 1977. The plant breeders' contribution to the origin and solution of pest and disease problems. pp. 71-81 in Origins of Pest, Parasite, Disease and Weed Problems. J. M. Cherrett and G. R. Sagar, eds. Blackwell Scientific Publications, Oxford.

Marchal, P. 1908. The utilization of auxiliary entomophagous insects in the struggle against insects injurious to agriculture. Pop. Sci. Monthly 72: 352-419.

NAS. 1971. Rapid Population Growth. Vols. I, II. Published for National Academy of Sciences by Johns Hopkins Press, Baltimore, Md.

NAS. 1975. Pest Control: An Assessment of Present and Alternative Technologies. Vol. II. Corn/Soybeans Pest Control. National Academy of Sciences, Washington, D.C.

NAS. 1977. Supporting Papers: World Food and Nutrition Study. Vol. 1. National Academy of Sciences, Washington, D.C. pp. 75-138.

NAS. 1978. Postharvest Food Losses in Developing Countries. BOSTID, National Academy of Sciences, Washington, D.C.

Oka, I. N., and D. Pimentel. 1976. Herbicide (2,4-D) increases insect and pathogen pests on corn. Science 193: 239-240.

PEP. 1955. World Population and Resources. Political and Economic Planning, London. 339 pp.

Pimentel, D. 1961a. Species diversity and insect population outbreaks. Ann. Entomol. Soc. Amer. 54: 76-86.

Pimentel, D. 1961b. The influence of plant spatial patterns on insect populations. Ann. Entomol. Soc. Amer. 54: 61-69.

Pimentel, D., and A. C. Bellotti. 1976. Parasite-host population systems and genetic stability. Am. Nat. 110: 877-888.

Pimentel, D., and M. Pimentel. 1979. Food, Energy and Society. Edward Arnold (Publishers) Ltd., London (in press).

Pimentel, D., W. Dritschilo, J. Krummel, and J. Kutzman. 1975. Energy and land constraints in food-protein production. Science 190: 754-761.

Root, R. B. 1973. Organization of a plant-arthropod association in simple and diverse habitats: the fauna of collards (B. oleracea). Ecol. Monogr. 43: 95-124.

Root, R. B. 1975. Some consequences of ecosystem texture. pp. 83-97 in Ecosystem Analysis and Prediction. S. A. Levin, ed. Soc. Ind. Appl. Math., Philadelphia.

Smith, R. F., T. E. Mittler, and C. N. Smith, eds. 1973. History of Entomology. Annual Reviews, Inc., Palo Alto.

Stern, V. M. 1969. Interplanting alfalfa in cotton to control lygus bugs and other insect pests. Proc. Tall Timbers Conf. Ecol. Anim. Contr. Habit. Mgmt. 1: 55-60.

Stevens, N. E., and W. O. Scott. 1950. How long will present spring oat varieties last in the central corn belt? Agron. J. 42: 307-309.

Tahvanainen, J. O., and R. B. Root. 1972. The influence of vegetational diversity on the population ecology of a specialized herbivore, Phyllotreta cruciferae (Coleoptera: Chrysomelidae). Oecologia (Berl.) 10: 321-346.

USDA. 1965. Losses in Agriculture. U.S. Dept. Agriculture. Agr. Handbook No. 291, Agr. Res. Serv., U.S. Govt. Print. Off., Washington, D.C.

USDA. 1977. Agricultural Statistics 1977. U.S. Dept. Agriculture. U.S. Govt. Print. Off., Washington, D.C.

van der Plank, J. E. 1968. Disease Resistance in Plants. Academic Press, New York. 206 pp.

Walker, J. C., R. H. Larson, and A. L. Taylor. 1958. Diseases of cabbage and related plants. USDA, Agr. Handbook No. 144.

2. The Quest for Innovation in Agricultural Entomology, 1945-1978

Abstract

New, synthetic, organic insecticides transformed insect-control practices after 1945. Serious problems surrounded the use of the new chemicals and included residues, resistance, induction of secondary-pest outbreaks, and concern over environmental hazards. The problems forced entomologists to search for new strategies of pest control that were less dependent upon chemicals. Two overlapping but in some ways rival paradigms for research emerged in the years following 1955. The pursuit of new control strategies by entomologists was influenced by philosophical assumptions, relations between entomologists and their client farmers, and efforts of entomologists to establish themselves as a strong profession. The dynamics of innovation in entomology have implications for public policy, which are briefly outlined.

Introduction

The insecticides with which Americans control pest insects have aroused more heated, emotional discussion than almost any other technology with the possible exception of nuclear power and birth control practices. Dozens of policy statements and studies, careful and otherwise, have been issued over the past thirty years by governments, universities, professional associations, industries, environmentalists, and other concerned individuals. Almost every conceivable position has at one time or another been defended. The range is from Wagnerian trumpetings about the essentiality of insecticides to prevent imminent famine and pestilence to virtual denials that they really are needed at all.

Most thoughtful analysts concluded long ago that extreme positions in this field were difficult to defend and that some uses of insecticides were to the benefit of all human

interests and others were not. There has also been virtual
agreement by all analysts that further research on insect
control was needed, and that, where possible, non-chemical
means of control should be fostered. Innovation has thus
been seen as the key to improved, non-chemical insect-control
practices. At the same time, surprisingly little attention
has been given to the processes by which professional
entomologists create the innovations everyone seems to want.
The general assumption has been that entomologists will respond "properly" if given sufficient money, research labs,
and general directions to improve the efficiency of control
techniques.

The tremendous impacts scientific and technological
innovation have had on our culture during the past one
hundred years have stimulated the development of an intellectual field concerned with relationships between scientific
creativity and the general cultural milieu. Some students of
this effort have argued that one cannot understand the origins
of scientific knowledge without understanding how the scientists are related to their own colleagues and to the many
types of people around them. This paper utilizes some of the
conceptual tools developed in the studies of science and
society to examine the research activities of entomologists in
the period of 1945-1978. The central concern is the question,
"What is the status of innovative activities designed to
create insect-control technologies less dependent upon chemicals?" The answers and their implications for public policy
are briefly outlined.

Changes in Entomology Before 1945

The debate over insecticides during the post-World War
II era occurred after a number of developments had already
taken place and irreversibly altered the stage of debate.
The most important of these included:

*The agricultural enterprise changed during the
nineteenth century from a subsistence way-of-life
into a commercial business. The importance of
insect control changed as a result. In subsistence
agriculture, the farmer's debt load was low and he
made few cash investments. Insect problems
caused losses of yield but did not threaten investments made with borrowed money; unless the outbreaks were catastrophic, the farmer bore them
without risk of losing his source of livelihood.
In commercialized agriculture, insect problems
threatened the safety of other cash investments
and thereby threatened the farmer's continued ability

to stay in business. The standards for acceptable
levels of insect control were thus higher than in
subsistence agriculture (Benedict, 1953, pp. 112-
122).

*Late in the nineteenth century entomology be-
came recognized as a distinctive, scientific
area. The social factor promoting the crystal-
lization of the field was the commercialization
of agriculture. The entomologists organized their
first national professional group, the American
Association for Economic Entomologists in 1889.
The establishment of the agricultural experiment
stations in 1888 led to many professional opportunities. By 1894, 42 States and Territories
employed persons for work on insects. In 1908,
the entomologists started a national journal for
publication of research, the Journal of Economic
Entomology (Howard, 1930, pp. 30, 69, 72-76, 106,
109).

*A variety of universities offered instruction
in entomology early in the nineteenth century,
but the establishment of the land-grant universities (1862) and the experiment stations vastly
increased the efforts in formal instruction. E.
Dwight Sanderson's Insect Pests of Farm, Garden
and Orchard (1915) provided a text for the instruction of new applied entomologists in the total
lore of the field (Flint and van den Bosch, 1977,
p. 102).

*The founding of the land grant universities and
the experiment stations established the role of
the public sector in research on insect control,
but the public sector also became involved in the
actual control of insects in the nineteenth
century. Massachusetts organized a commission
to exterminate the gypsy moth in 1890, and it
carried out extensive control operations on private
lands until the legislature declined to continue
its support in 1900 (Forbush and Fernald, 1896;
Dunlap, 1978).

*The attractiveness of insecticides in allowing
individual farmers to make pest control decisions
was recognized in the nineteenth century and
earlier. The arsenicals such as Paris green,
lead arsenate, and calcium arsenate, plus botanical
insecticides such as pyrethrum, rotenone, and

others were widely used by 1930. In addition,
an extensive controversy over the safety of residues
of insecticides brought farmers and the government
repeatedly to heated litigation before 1940
(Shepard, 1951; Whorton, 1974).

The major features of disputes about insecticides were
thus all present in America before 1945: the financial
pressures of commercial agriculture, a dynamic community
of professional entomologists employed largely in the
public-sector, government sponsored control programs,
insecticides, and disputes over the safety of the chemicals.
The controversies changed and grew after 1945, but they were
clearly grounded in trends and traditions that had emerged
much earlier.

Changes in Entomology (1945-1978)

Agricultural entomology was marked by three overlapping
periods between 1945 and 1978. The first, Euphoria and the
Crisis of Residues, occupied the period from about 1945 to
about 1955. The second, Confusion and the Crisis of
Environment, fell in the period from 1954 to about 1972.
The third and current period, Changing Paradigms, began in
about 1968 and has not yet ended.

Euphoria and the Crisis of Residues (1945-1955)

The chemist, Paul Herman Müller, of the J. R. Geigy Co.,
Switzerland, discovered the insecticidal properties of DDT
in 1939 and thereby initiated a host of important changes in
entomology. DDT was successful commercially and, more importantly, its successes indicated that the organic chemist
could hope to identify additional insecticides amidst the
nearly infinite number of molecules that could be synthesized.
The chemists were successful, and in the ten years after
World War II, many companies introduced a wide variety of new
products such as benzene hexachloride, toxaphene, chlordane,
parathion, and others. Their adoption in agriculture was
rapid and euphoric (Perkins, 1978b).

The source of euphoria was simple: the new, synthetic
organic insecticides allowed entomologists and their client
farmers to achieve a degree of insect control that was simply
unheard of before their introduction. Calcium arsenate, lead
arsenate, botanical insecticides, and other materials used
before 1939 could not compete with the new "miracle chemicals." The reports of spectacular successes came from all
agricultural areas. Apple growers in the Yakima Valley of
Washington saw their losses from codling moth drop from 15%

with lead arsenate to 3-5% with DDT. Potato growers found
that DDT helped increase their yields from 155 bushels per
acre (1945) to 211 bushels per acre (1949) (Bishopp, 1951,
pp. 376, 378). Cotton growers in Louisiana, heavily plagued
with the boll weevil since the first decade of the century,
found that toxaphene and other new insecticides controlled
the weevil far better than calcium arsenate, the only
material with even moderate effectiveness up to that time
(Newsom, 1974; NAS, 1975b).

All across the nation, farmers and entomologists alike
scrambled to exploit the technological power of the new
insecticides. Neither group had much choice in the matter.
Farmers who refused to adopt profitable techniques risked
being forced out of business by their technologically more
progressive neighbors. Scientists who preferred other lines
of research risked professional stagnation by not investi-
gating what were on the surface, the most marked advances in
insect control ever demonstrated. Some entomologists enthusiastic about biological control, for example, later recalled that they were ridiculed as a lunatic fringe by their
chemically inclined colleagues (Doutt and Smith, 1971).

The development of DDT had impacts on all aspects of
entomological knowledge:

> Insect-control research and practices in the
> United States were thus reshaped. . . : (1) the
> success of DDT stimulated the development of other
> synthetic organic insecticides, (2) old chemicals
> were abandoned for new ones, (3) chemical control
> technologies acquired a greater prominence in the
> total constellation of insect-control technologies,
> (4) biological control technologies were disrupted,
> (5) control practices based on habitat sanitation
> and cultural practices were abandoned, (6) eradi-
> cation proposals won new adherents, and (7) research
> problems undertaken by entomologists shifted from
> biological studies toward studies of insecticides.
> (Perkins, 1978b).

The euphoria among entomologists and farmers over the
new insecticides was not, however, untouched by other concerns. Even before DDT was released by the War Production
Board for civilian use, concerns were voiced by naturalists
over the safety of the compound for widespread use in the
environment (Conant, 1944). These concerns were picked up
and amplified by other wildlife biologists once DDT began to
receive widespread use, especially for control of forest
insect pests (Cottam, 1946). The evidence for hazard from

doses likely to be used on forests, however, was mixed (Brues, 1947; Kendeigh, 1947).

The concern that caught wide public attention, however, was not the potential for damage to the environment. Rather, in the late 1940's increasing numbers of questions were raised about the safety of the new compounds to humans. Were the minute quantities of residues left on treated agricultural crops really safe, even over a long period of time? Were hazards increased for infants, pregnant women, older people, and people with weakening conditions? Entomologists were aware of these questions, but they had insufficient data with which to answer the critics. The residue questions sparked the first post-war crisis for the science.

Commissioner Paul Dunbar of the Food and Drug Administration (FDA) announced in 1949 a series of hearings on residues of insecticides at the same time Representative Frank Keefe (R., Wisc.) introduced legislation establishing the Select Committee on Chemical in Foods and Cosmetics (Federal Register, 1949; U. S. Congress, 1950). The FDA hearings were narrow in scope. The government wanted to know what fruits and vegetables required the use of poisons in their production, what poisons were required, how much poison could be removed before marketing, and the quantity that could be tolerated without endangering the public health. The mission of the Select Committee (the Delaney Committee after its chairman James J. Delaney (D., N. Y.)) was considerably broader in scope than the FDA hearings because it included pesticides, fertilizers, food preservatives, food additives, packaging materials, and chemicals found in foods and cosmetics. In the shortrun, the FDA provided the larger threat to the use of insecticides because the FDA already had the power to make rules that might drastically affect farmers' abilities to use them. In the long-run, the Delaney Committee was more important because they might recommend sweeping new legislation of even less predictable impact than tolerances set by the FDA.

Entomologists Sievert A. Rohwer and Fred C. Bishopp of the Bureau of Entomology and Plant Quarantine, USDA, played leading roles in providing the FDA and the Delaney Committee with needed technical expertise. Their judgments were supplemented by scientists from the land grant universities.

Bishopp's testimony to the Delaney Committee reflected the conservative stance taken by both federal and state entomologists toward outside intervention. He argued that (a) present laws were adequate to protect the public health, (b) DDT as the major example of the new type of insecticide had been tremendously successful in cutting losses, and (c) there was no evidence that the toxic hazards of DDT had been underestimated. Bishopp was in emphatic disagreement with contrary testimony presented by Arnold J. Lehman of the FDA (Bishopp, 1951, pp. 376-380, 387, 400).

The Committee concluded that insecticides then in use frequently had insufficient toxicological or pharmacological information to establish their safety and that new laws were needed (U. S. Congress, 1952). Congressman-physician A. L. Miller (R., Neb.), a member of the Delaney Committee, introduced an amendment to the Food, Drug and Cosmetic Act that required pesticide manufacturers to test their products for human safety and obtain a tolerance or exemption from a tolerance before marketing it. Congress passed the Miller Amendment in 1954 (68 Stat. 511-517) and mooted the hearings of the FDA by specifying a new method for setting tolerances.

The professional pride of entomology was wounded by the FDA hearings and the work of the Delaney Committee. Outsiders had raised questions about the prize chemicals of the entomologists, and a new law was enacted to regulate them. Entomologist E. F. Knipling, President of The American Association for Economic Entomology, remarked in late 1952 that entomologists had not done enough in telling the public why insecticides were important. He agreed more toxicological data were needed, but he believed the Congress had received distorted information (Knipling, 1953). The wounding of professional pride, however, did not substantially affect the professional autonomy of entomologists. The chemical manufacturers now had to obtain a tolerance before they registered a new insecticide, but the science of entomology itself was unaffected for all practical purposes.

Confusion and the Crisis of the Environment (1954-1972)

The second period was more complex than the first, and a multitude of factors shaped it. The most important were increased use of capital in farming, shifts in the biological properties of insect populations, a series of conceptual and technical developments in the discipline itself, and another serious intervention from the outside. The kaleidoscope of changing conditions left the science in disarray until the late 1960's.

The commercial farming business continued its shift toward the increased use of capital which had begun much earlier (Heady, 1967, pp. 15-19). The number of farms and people engaged in agriculture decreased; the size of farms increased; and productivity per acre and per person-hour increased. All

of these changes were correlated with changing technologies
including those for insect control. Entomological developments, particularly the new insecticides, were partially responsible for the switch to capital-intensive agriculture.
Once the switch was made, however, entomology faced a new
problem. The scientists were under increased pressure to
maintain the level of control their innovations provided
because the farmer could no longer return to more labor-
intensive production. The initial successes of the new
insecticides thus created a condition that demanded their
continued effectiveness.

The changing milieu of agriculture was particularly
evident in corn and cotton production. The western corn
rootworm was one of several insects that could damage corn
in the Midwest. Prior to the introduction of the synthetic
organic insecticides, it was often controlled by crop rotations. The chlorinated hydrocarbons DDT, benzene hexachloride, and later aldrin, however, allowed the corn farmer to
stop rotations, treat his soil with the chemical, and
thereby grow corn year after year without risk of a serious
rootworm outbreak (Perkins, 1978b). In a similar vein, the
cotton boll weevil had plagued cotton from Virginia to Texas
since 1922. Calcium arsenate had provided some relief, but
the use of early planting, short season varieties, early
harvesting, and stalk destruction were the core of control
methods recommended by entomologists (Helms, 1979). The
new insecticides, especially benzene hexachloride and toxaphene, allowed growers to adopt longer season varieties,
irrigation, and heavy fertilization, all of which aggravated
the boll weevil problem (Newsom, 1974). Many corn and cotton
farmers in the late 1950's, therefore, were using production
systems that had been partially changed by the entomologists
and which in turn required the entomologists to ensure the
continued success of their methods.

Control schemes based on chemicals were undermined by
changes in the properties of insect populations. The development of resistance to insecticides was one of the most
serious such developments. Between 1955 and 1960, the
phenomenon forced the profession to re-examine its strategy
of research and development. The reports of resistance
between 1955 and 1960 were by no means the first. Resistance had been reported first in 1908 when the San Jose scale
refused to respond to treatment with lime sulfur. By 1945,
thirteen species had been reported as showing some resistance
to insecticides. The widespread deployment of the new
insecticides was followed almost immediately by a jump in
the number of species escaping the selective pressure of
insecticide. The resistance problem began to arouse serious

concern among entomologists in the late 1940's. In 1960,
Anthony W. A. Brown delivered a major address on the subject
to the Entomological Society of America and documented its
occurrence in 124 new species since 1945. He concluded the
golden age of insect control by chemicals had already
passed (Brown, 1961).

The second change in insects was the widespread destruction by the new insecticides of beneficial insects
including predators and parasites and also pollinating
species and honeybees. The most serious results were resurgence (flareback) and the creation of secondary-pest
outbreaks. Both rest on a common basis. Most plant-eating
insects have predatory and parasitic insects (beneficial
natural enemies) that keep their population densities low.
If these beneficial species are destroyed by insecticides,
the pest insect may be able to rebuild its population more
rapidly (resurge or flareback) than if the natural enemies
had been present. It is also possible that a pest insect
formerly kept in control by natural enemies may rise in
density to pest status if natural enemies are destroyed by
insecticides. Such an insect is said to be a secondary pest.
The phenomena of rèsurgence and the creation of secondary-
pest outbreaks had occurred with materials used before the
advent of DDT, but the magnitudes of the problems were so
increased by the new insecticides that they could no longer
be ignored (Ripper, 1956).

The technological failure of insecticides due to resistance, resurgence, and induction of secondary pests
reached a crisis stage in cotton in the late 1960's. The
cotton bollworm and the tobacco budworm, both secondary
pests on cotton, became resistant to the chlorinated hydrocarbon insecticides, the carbamates, and, in the budworm,
the organophosphates. Secondary insect pests that had been
induced by insecticides thus were no longer controlled by
them (NAS, 1975b). The situation created near panic among
entomologists and farmers alike in southern Texas where the
situation was most grim. Entomologist Perry Lee Adkisson
of Texas A&M University still remembers those days of the
late 1960's as a severe trial by fire for the entomological
profession as a whole and for his department in particular
(Perry L. Adkisson, Personal Interview, 1978).

The resolution of the problems and crises associated
with insecticides beginning in the 1950's came from a
revival of research avenues known long before DDT plus new
discoveries and developments. The dramatic successes of the
new insecticides had eclipsed the importance of the older
entomological practices, but the research traditions on which

they were based had never totally disappeared. Examples of some of the developments based on older research avenues plus exciting new discoveries are listed below. Many date from before the 1940's, but by 1960 they came to be seen in a new light.

*1939-1950: Ralph T. White and Samson R. Dutky (USDA) developed and disseminated the milky disease for biological control of the Japanese beetle on 93,000 acres in the Eastern United States (Hawley, 1952; Dutky, 1952).

*1945-1950: James K. Holloway (USDA) and Carl B. Huffaker (University of California) achieved biological control of the Klamath weed in California with insect enemies (Holloway and Huffaker, 1952).

*1951-1958: Edward A. Steinhaus (University of California), demonstrated Bacillus thuringiensis could reduce populations of alfalfa caterpillar; the first commercial preparations of Bacillus thuringiensis became available for testing by entomologists in 1958 (Steinhaus, 1951; Hall, 1963; Heimpel and Angus, 1963).

*1932-1943: Reginald H. Painter (Kansas State University), and other colleagues in Kansas, Nebraska, USDA and elsewhere developed and introduced Pawnee wheat, resistant to the Hessian fly; by 1946 this variety of wheat was the leading one grown in central and eastern Nebraska (Painter, 1951, pp. 148-153).

*1937-1959: Edward F. Knipling, Raymond C. Bushland, and other USDA colleagues developed the sterile-male technique and eradicated the screwworm fly from Curacao and Florida (Knipling, 1959; Scruggs, 1975; Perkins, 1978a).

*1956-1960: Leo Dale Newsom and James Roland Brazzel (Louisiana State University), identified diapause in the boll weevil (Brazzel and Newsom, 1959). Brazzel (then at Texas A&M University), Theodore B. Davich (USDA), and other colleagues translated this observation into a new way to use insecticides, the diapause control method by 1960 (Brazzel, Davich, and Harris, 1961).

*1917-1956: Carroll M. Williams (Harvard Uni-
versity) and many others established that a
sequence of hormones regulates growth and
metamorphosis of insects. Williams reported
the first extract of juvenile hormone in
1956 (Meyer, 1972; Steele, 1976; Wigglesworth,
1970).

*1959: Peter Karlson and Adolf Butenandt
(Max-Planck-Institut für Biochemie) coin
the term "pheromone" and thus crystalize
the long-developing field of insect attractants into a new sub-discipline of entomology
(Karlson and Butenandt, 1959).

While the entomologists were reviving old lines of
research and developed new possibilities, the chemical
industry continued to introduce new chemicals. Union
Carbide's carbaryl (1958), Farbenfabriken Bayer's methyl
parathion (1952), and Geigy's diazinon (1958) are only
three examples of materials that quickly achieved wide
acceptance. It is important to note here that the threats
posed by resistance, resurgence, and secondary pests did not
slacken the chemical industry's desire or ability to introduce new compounds with a broad spectrum of toxicity. From
the industry's point of view, in fact, a new chemical was
the way to solve a problem created by an old one.

The emergence of so many potential new avenues of research was stimulating, but the very plethora of possibilities created its own problems. Which paths of research
were most likely to be successful? Should entomologists
pursue all of them with equal vigor, or should priorities
be made. If choices were to be made, how, by whom, and with
what criteria?

Gradually during the 1950's and early 1960's,
entomologists from the USDA and the land grant universities
began to search for alternatives to the use of chemicals
that would provide adequate levels of insect control without
the disadvantages of the chemicals. In addition to the
search for alternatives, entomologists began to grapple with
a related but more difficult problem: how to combine two or
more control technologies into programs in which each
individual method could synergize the effectiveness of
others and thus create a level of suppression greater than
that provided by a single technique.

The reorientation of research towards non-chemical
control methods was clearly underway by 1960 because of

actions taken by many leaders within the profession. It is
difficult to identify precisely the contributions of individuals, but Edward F. Knipling, Chief of the Entomological
Research Division at USDA, must be noted as one person who
contributed disproportionately. One reason for his importance, of course, was that he was responsible for directing
the largest entomological research institution in the U. S.
Any changes in USDA research policy automatically influenced
the total constellation of research activities within the
country. Only a few land grant universities were sufficiently large that changes made by them resulted in a significant
difference in the overall picture. More important than
Knipling's administrative responsibility, however, was the
fact that he was one of the first individuals to speak
strongly for new concepts of entomological research <u>on a</u>
<u>systematic basis</u>.

The reorientation of entomological research did not
occur instantly; indeed decades were the time unit needed to
see the process. Furthermore, as will be discussed in the
next section, the processes of reorientation were marked by
the emergence of different visions about the goals that
entomologists should entertain.

The reorientation of entomological research began as an
internal movement within the profession, but it did not remain so for long. While entomologists were searching for
new concepts on which to base their research and recommendations, Rachel Carson launched a broadside against the widespread use of insecticides. Her well-known <u>Silent Spring</u>
(Carson, 1962) appeared in 1962, and the ensuing furor took
the matter of insect control all the way to the White House
and President John F. Kennedy.

Carson's theme was that pesticides were used carelessly
and without sophistication in ways that created hazards for
non-target organisms and resulted in the technological
failure of resistance, resurgence, and secondary pests. All
of these problems had been identified prior to Carson's work,
and entomologists who read outside their narrow specialties
were probably aware of most if not all of them. Carson's
unique contribution, however, was a synthesis of widely
scattered literature with an emphasis on contamination of the
environment.* She also summarized a large number of

---

* Brooks (1972, pp. 214-216) records that Carson herself believed she was preparing a unique synthesis of widely scattered literature. Rudd (1964) worked simultaneously with
Carson on similar but more technical synthesis, but his
book was published over a year after Carson's.

alternative avenues of productive research. Implicit in her
work was the analytical framework now accepted for risk/
benefit evaluations of pest control practices.

The responses to Silent Spring ranged from vitriolic to
laudatory. The most widely circulated public statements
critical of it came not from the entomological community
but from scientists dealing with health and food processing.
This is not to say that entomologists did not grumble about
it privately, but the published literature contains only a
few of their comments on it.* In a press release, the USDA
acknowledged Carson's work to be ". . . a lucid description
of the real and potential dangers of misusing chemical
pesticides" (USDA, 1962). After granting the correctness of
her thesis, the Department went on to say that they knew
about the problems and had already taken steps to correct
them. The reorientation of insect control research that had
begun in USDA upon Knipling's appointment in 1953 as chief
of entomological research supported the Department's statement.

Had Carson not written her book, there was no indication
that the arguments about insect control would ever have
reached beyond a handful of farm interests, chemical companies, wildlife specialists, professional entomologists,
USDA and University administrators, and a few Congressmen.
Silent Spring, however, dramatically changed the politics of
the debate.

John F. Kennedy directed his President's Science
Advisory Committee (PSAC) to examine the issues, and their
report of May 15, 1963, vindicated Carson's major points
about the need for non-chemical methods of pest control
(PSAC, 1963). Senator Abraham Ribicoff (D., Ct.) then
launched an extensive series of hearings on pest control
(U. S. Congress, 1966b). PSAC recommended non-chemical

---

* The most widely circulated negative reviews were I. L.
Baldwin (1962) and William J. Darby (1962). Book Review
Digest for 1962 and 1963 records 20 reviews of which 13 were
positive, 3 were intermediate, 3 were unclassified, and 1
(Darby's) was negative. The only formal review of the book
by an entomologist I am aware of is P. J. Chapman (1963).
(I thank David Pimentel for bringing this review to my attention). Chapman was intermediate in his praise and criticism.
Cornell entomologist E. H. Smith (1964) briefly touched on
her book but gave no formal criticisms other than that Carson
had been emotional and unobjective. The Bulletin of the
Entomological Society of America did not review the book,
unfortunately, although they did review Rudd (1964).

approaches to pest control, but it proved difficult to implement the recommendation. The Senate produced minor legislation providing that pesticides could no longer be registered by a manufacturer against the wishes of the government and that the registration number could be shown on the label (78 Stat. 190). The immediate effects of the PSAC report and the Senate hearings were thus modest in the short run, but they were important in legitimating the debate on insecticides among a wide number of interested parties.

Carson's book in the short run helped stimulate an increase in funds for entomological research. The budget for research in USDA's Entomology Research Division rose from $11.2 million in 1964 to $16.9 million in 1965 (U. S. Congress, 1965, pp. 737-738; 1966a, pp. 533-534). This was the largest jump in funding the Division had received in one year since World War II. For all the professional dignity tarnished by Carson, the entomological research enterprise benefitted.*

In addition, Carson's book stimulated an organized response from a number of state and federal scientists concerned with pests and pesticides. The Federal Committee on Pest Control, responsible for monitoring and coordinating federal activities related to pests, initiated a symposium at the National Academy of Sciences, held in 1966. The thirty authors presenting twenty-eight papers included zoologists, entomologists, toxicologists, medical scientists, chemists, industrialists, a geneticist, a Congressman, and the Secretary of the Interior (NAS, 1966, pp. x, 368.

The broad spectrum of opinion and expertise represented in part an effort by scientists closest to the pesticide situation to demonstrate their broad competence and concern for the multitude of social and technical problems surrounding pest control. For some participants, the Symposium was an opportunity to discuss environmental and health hazards of pesticides in scientific circles that previous to Carson's book had been unreceptive to such material. For others, the event was an opportunity to provide "scientific" balance to

---

* Fish-kills in the lower Mississippi River were also powerful motivating forces. Hearings on the episode were held in April and May, 1964. Hearings for the fiscal year 1965 budget had already been completed, but the House and Senate Committees on Appropriations both increased the research allotments for pest control as a result of the incidents. (U. S. Congress, 1965, pp. 749-752, 826-827, 856-858).

Carson's approach, which they considered seriously biased.
As with the PSAC report and the Senate hearings, the immediate impact of the proceedings, Scientific Aspects of Pest
Control, were difficult to see and undoubtedly modest. Over
a longer period, the report enlarged the boundaries of
legitimate concerns in insect control science.

Silent Spring thus generated a flurry of scientific and
political activity, but it did not end the rapid growth of
the synthetic insecticides industry which enjoyed a continued
rise in sales during the 1960's. Two important institutional
changes, however, occurred by 1972. The Nixon Administration
transferred pesticide regulation from the USDA to the newly
formed Environmental Protection Agency in 1970 and thereby
removed insecticide governance from an agency that had been
reluctant to prohibit the use of any chemical. As a result,
pressures brought by widespread press coverage of pesticide
problems, legal actions by groups such as the Environmental
Defense Fund, and by concerns voiced in the Department of
Health, Education, and Welfare found a more sympathetic ear.
DDT, chlordane, heptachlor, aldrin, dieldrin, endrin, and
others were totally or partially deregistered by 1977 (USEPA,
1977). In addition, a new pesticide law, the Federal Environmental Pesticide Control Act, was passed by the Congress
in 1972 (86 Stat. 973-999). This new legislation formalized
protection of the environment as public policy. The disputes
about insect control did not end, but the politics were
changed.

Changing Paradigms (1968-present)

Thomas S. Kuhn made the term "paradigm" a highly popular
one for analysts of scientific and technical change. His
book, The Structure of Scientific Revolutions (Kuhn, 1970),
presents a general model for scientific change based upon
the emergence, use, and eventual discard of paradigms in
scientific communities. The criticisms and responses to
Kuhn's work (Lakatos and Musgrave, 1970) suggest that his
theory is not without difficulties that are beyond the scope
of this paper to review. As a first approximation, however,
his notion of "paradigm shift" as the unit of scientific
change helps order the events in entomology that have
occurred since 1945.

Kuhn believes that most scientific work, or normal
science, is performed by a practitioner working with a
particular paradigm. Paradigms consist of the results or
exemplars of past work that are accepted by a community of
scientists and that supply the foundation for their further
work. The two essential characteristics of paradigms are

(a) their ability to attract an enduring group of adherents away from competing modes of activity, and (b) the presence of a sufficient number of problems for the adherents to resolve (Kuhn, 1970, p. 10).

In Kuhn's language, the events between 1945 and the mid-1950's can be summarized as follows: The new synthetic organic insecticides allowed the development of a new paradigm for applied entomology: the major tool for controlling insects would be the application of toxic chemicals to them. Other methods were not forgotten, left totally unused, or dismissed from all research efforts; but they were for the most part relegated to secondary importance. This chemical-control paradigm attracted many adherents away from competing lines of research and provided numerous problems of normal science for entomologists. Particularly important were the questions of which chemical, applied how, at what times, and in what amounts. When asked, entomologists would maintain that insects could be controlled by many different means; when drawing up their own research plans, the many entomologists who adhered to the chemical-control paradigm selected a chemical as the foundation of the experimental design.

The published research literature reflected the dominance of the chemical-control paradigm: British entomologist, D. Price Jones surveyed the Journal of Economic Entomology between 1927 and 1970 and found that the percentage of articles devoted to chemical-control rose sharply in the late 1930's and especially after 1945 until the late 1950's (Jones, 1973). This record in the major journal of the entomological community speaks eloquently for the power of the chemical-control paradigm in shaping research in the period 1945-1960.

Two major, and in some ways alternative, new paradigms were developed by small groups of entomologists in the years following 1955. The two shared many characteristics, but the differences between them became sufficiently great that they must now be distinguished. Both paradigms were developed because of the problems associated with heavy reliance on insecticides, especially resistance and destruction of natural enemies. Neither totally rejected the use of chemicals, and both sought ways in which the powerful benefits of insecticides could be obtained without their problems and risks. Both recognized the need for a package of different control techniques and a systematic method for integrating them. The fundamental difference between them centered on the ultimate goal toward which entomological research should be aimed.

The significant overlaps between the two paradigms
created in the entomological literature a certain amount of
confusion in that both utilized terms such as "pest management," "integrated pest management," "integrated control,"
and others. In this analysis, I will use the terms "Integrated Pest Management" (IPM) and "Total Population Management" (TPM) to designate the alternative paradigms. The
term IPM was coined and used by its proponents, but TPM is
my term derived from language used by its proponents. In
the entomological literature, the term "integrated" is used
by both schools of thought.

It should also be understood that the analysis presented
here is neither criticism nor advocacy of one compared to the
other on either biological or social grounds. Such an
analysis is beyond the scope of this paper but will be presented elsewhere.

The theory and vision of what is now called Integrated
Pest Management were developed over a period of decades by
a number of entomologists in the U.S., Canada, and elsewhere.
Especially prominent early spokesmen and theoreticians
included A. E. Michelbacher (University of California), A. D.
Pickett (Dominion Entomological Laboratory, Canada), Ray F.
Smith (University of California), and others. Michelbacher
appears to be the first to have used the term "integrated
control" in 1952 (Michelbacher and Bacon, 1952).[*] Pickett
developed serious reservations over what he saw as careless
uses of insecticides and wrote an eloquent call for ecologically based pest control in 1949 (Pickett, 1949). Smith
(1975) argues that the origin of IPM in terms of philosophy
and practice dated to the late nineteenth century and the
works of Charles W. Woodworth (University of California).
Many of the major theoreticians of the IPM school were
located at the University of California and heavily influenced by the philosophies, methods, and successes of
"classical biological control": the search of foreign lands
for predatory and parasitic insects followed by their im-
portation, release, and evaluation as control agents in the
U.S. Accordingly, the first formal articulation of the
integrated-control paradigm in 1959 referred to the integration of only two techniques: biological and chemical
(Stern, et al., 1959). By the early 1970's, the paradigm
was more fully developed and called for the integration of

---

[*] I thank Kenneth S. Hagen for bringing this article to my
attention.

all feasible control technologies guided by an understanding of the ecology of the agricultural system.*

The IPM paradigm also included from 1959 the notions of "economic thresholds" and the need for "supervised control;" both reflect social aspects of pest control rather than biological ones. The economic threshold was the population density of the pest species at which the cost of a control effort was repaid by the value of damages prevented. Pest populations less dense than the economic threshold were, in the IPM paradigm, not worth treating because the cost of treatment would be more than the value of damage prevented. Supervised control referred to the concept that considerable expertise was needed to make proper decisions in the field. IPM proponents argued that farmers would be unlikely to have the expertise needed and would therefore need assistance from professional entomologists.

The developers and proponents of IPM have offered a variety of definitions of their paradigm. One of the more comprehensive was offered in 1971 by researchers from the University of California:

> Integrated control is a pest population manage-
> ment system that utilizes all suitable techniques
> either to reduce pest populations and maintain
> them at levels below those causing economic
> injury, or to so manipulate the population
> that they are prevented from causing such injury.
> Integrated control achieves this ideal by har-
> monizing techniques in an organized way, by
> making the techniques compatible, and by blending
> them into a multifaceted, flexible system . . .
> In other words, it is an holistic approach aimed
> at minimizing pest impact while simultaneously
> maintaining the integrity of the ecosystem
> (Corbet and Smith, 1976, p. 662).

The IPM paradigm inspired and guided a wide variety of research projects in both federal and state laboratories including those utilizing classical biological control,

---

* An FAO Symposium on Integrated Control was held in Rome in September, 1965. It provided the impetus to "crystallize" the integrated control concept as one encompassing all techniques (Kennedy, 1968). Ray F. Smith and Hal T. Reynolds offered the first definition of the more inclusive concept at that time (Smith and Reynolds, 1966). See R. F. Smith and R. van den Bosch (1967) for an eloquent vision of the fundamental parameters of agroecosystems.

cultural control, insecticides, host plant resistance, and, in the 1970's, agroecosystem modelling with the aid of systems analysis and computers.

The theory and vision of Total Population Management (TPM) (my term) was articulated primarily by Edward F. Knipling (USDA), but it attracted significant attention from other entomologists as well as political figures and prominent members of the farming industries. The TPM-paradigm was developed after 1955 and underwent continued refinements into the 1970's. It was sufficiently mature by 1965 to recognize it as distinct from the IPM-school. Knipling argued in the Founder's Memorial Lecture (a major honor) to the Entomological Society of America in late 1965 that for some key insect pests the proper target for control was the total population of the species over a significant geographic area. Furthermore, eradication of certain key pests might be achieved with TPM. Even if eradication were unsuccessful, it was worthy of consideration because, Knipling argued, even a few successes would return immense benefits to society in the form of reduced losses and de-creased environmental contamination from the use of insecticides (Knipling, 1966b). Eradication was not en-visioned in the IPM-paradigm.

It is important to understand that Knipling and others attracted to the TPM paradigm did not reject the IPM-paradigm. Rather, Knipling viewed the concept of TPM as a progressive step beyond IPM and justified only for a few species. He also believed that control techniques developed under the IPM-paradigm were a "fall-back" position from any eradication effort that did not succeed. In his view, re-search inspired and guided by the TPM-paradigm was also use-ful as a basis for control practices under the IPM-paradigm (E. F. Knipling, Personal Communication, 1978).

The TPM-paradigm inspired a wide variety of research activities. As noted earlier, Knipling, working with his Branch Chiefs, began to shift federal entomological research into non-chemical control projects after his appointment as chief of entomological research in USDA in 1953. Between 1953 and the mid-1960's federal research dollars moved from an estimated two thirds on chemical control to about 16% on chemicals (Hoffmann, 1970).* Knipling's position as director of entomological research in USDA gave him consider-able influence in conceptualizing and implementing

---

* I thank E. F. Knipling for bringing this article to my attention.

coordinated research efforts within the TPM-paradigm. An
excellent example of such a research package was that carried
on at the Boll Weevil Research Laboratory in Mississippi
under the direction of Theodore B. Davich.

The Boll Weevil Research Laboratory (BWRL) was established in 1961 after the cotton industry had requested help
from the Congress for relief from the production losses
caused by the boll weevil. The BWRL's multidisciplinary
team of researchers developed a wide range of projects in
the 1960's including work on basic ecology of the insect,
pheromone attractants, host plant resistance, improved
methods of chemical control, basic physiology of the cotton
plant, feeding stimulants and inhibitors for the boll weevil,
methods of mass-rearing and sterilization, and others
(Agricultural Research Service, 1962; T. B. Davich, Personal
Interview, 1978; E. F. Knipling, Personal Communication,
1978).

The research package developed at the BWRL was heavily
influenced by Knipling's TPM-paradigm. At the dedication of
the Laboratory in 1962, he said:

> . . . Congress expects more than minor improve-
> ments. . . . Therefore, the objective of the re-
> search should be to find ways of reducing losses
> to a minimum or to eliminate the problem entirely.
> For my part, I feel that we should gear our
> thinking and direct our research efforts to the
> development of practical ways of eradicating the
> insect. I am confident that research workers
> can achieve this objective . . . Now it is my
> view, based on a great deal of thought and
> study of the population dynamics of insects,
> that the difference between a high degree of
> control of an insect like the boll weevil and
> complete elimination of the pest is a rather
> narrow one in terms of the actual number of
> insects involved (Knipling, 1962, p. 2).

Even though Knipling clearly expressed his notion of
how the research at the BWRL should be directed towards
eradication, one possible goal contained within the TPM-
paradigm, some of the research was equally well suited to
non-eradication control measures.

In 1968, entomologists Knipling, James Brazzel (USDA),
Theodore Davich (USDA), Perry Adkisson (Texas A&M
University), David Young (Mississippi State University), and
C. R. Jordan (University of Georgia) served on the National

Cotton Council's Special Study Committee on Boll Weevil
Eradication. This group conceptualized a multi-million
dollar experiment (the Pilot Boll Weevil Eradication
Experiment or PBWEE) to test whether technology was adequate
to eradicate the boll weevil (Special Study Committee on
Boll Weevil Eradication, 1969). It was the largest and
most complex exercise ever attempted in entomology research
and it was the first test of the TPM-paradigm on a large
scale. Only the earlier work leading to the eradication of
the screwworm fly from Florida (1958-1959) and its suppression in the Southwest (1962 and after) were comparable and
served as inspiration for the PBWEE. Strong proponents of
the TPM school viewed the eradication experiment as a progressive step beyond research in the IPM-school, but they
did not reject IPM-research as either unworkable or unworthy. In their vision, TPM led them to do research on
more powerful techniques they believed could and should be
done in efforts to deal with the boll weevil (Perkins, in
preparation).

Carl Barton Huffaker's (University of California)
connections with the International Biological Program in the
mid to late 1960's led to a series of events that resulted
in a second multi-million dollar research exercise during
1972-1977: "The Principles, Strategies and Tactics of Pest
Population Regulation and Control in Major Crop Ecosystems,"
more commonly known as the "Huffaker Project" because of
Huffaker's role as director. The Huffaker project was a
coordinated research effort between 19 land-grant universities and the USDA on six major agroecosystems: cotton,
soybeans, stone and pome fruits, citrus, alfalfa, and pine
forests.* The philosophy of the Huffaker Project was clearly
derived from the IPM school, and the leadership of the project, especially Huffaker and Ray F. Smith (University of
California) were disenchanted with the TPM-paradigm in
general and eradication efforts in particular (Carl B.
Huffaker, Personal Interview, 1977; Smith, Apple, and
Bottrell, 1976, p. 11; Apple and Smith, 1976, pp. 184-186).
Other entomologists active in the Huffaker Project may have
held considerable interest in the TPM paradigm, but the research of the Huffaker Project itself did not include
efforts toward eradication.

The Huffaker Project and the PBWEE were not strictly
comparable because the former was primarily a research

---

* Somewhat similar independent research was carried on at
the same time in USDA and other state experiment stations;
State and Federal extension workers also began IPM education/
demonstration projects in 1972.

project with some field demonstrations but the latter was
largely a field demonstration supported by a research effort.
The two exercises were comparable, however, on two funda-
mentally important points: (a) they both required and received significant political support, and (b) they were both
manifestations of underlying philosophical perspectives that
were in some ways rivals. As there were no other comparable,
major exercises in entomology at this time, the antagonisms
between the two schools of thought (TPM and IPM) were frequently expressed in reference to these two projects.

The results of the PBWEE experiment (conducted between
(1971 and 1973) were ambiguous in the sense that a few boll
weevils were found in the eradication zone at the conclusion
of the effort (Entomological Society of America Review
Committee, 1973). Strong proponents of the TPM paradigm
(Knipling, Brazzel, Davich, and others) regarded the results
as extraordinarily encouraging. They argued that the
technology then in hand, plus refinements that would come
through further research were adequate to start an eradication program against the weevil from Virginia to Texas.
Furthermore, even if eradication failed, it was worth the
gamble (Perkins, in preparation; Knipling, Personal Inter-
viw, 1976; Knipling, 1978). Other entomologists (such as
Adkisson, Smith, Huffaker, and others) viewed the boll
weevils remaining in the eradication zone negatively. They
argued that the technology for eradication was not then
available and that, in their judgment, further refinements
either would not develop it or even if developed would not
be cost effective. Instead, adherents of the IPM-school
argued that the best solution to the boll weevil problem was
to be found in research of the type that had been conducted
under the umbrella of the IPM paradigm (Perry Adkisson,
Personal Interview, 1978; Carl Huffaker, Personal Interview,
1977; Perkins, in preparation).

The ability to distinguish the two competing paradigms
and particular individuals attached to one or the other does
not imply that entomologists were easily divided into two
opposing camps. The sharing of some common features between
IPM and TPM plus the fact that both were in process of
development and maturation into the 1970's, allowed some
entomologists to perceive no overwhelming conflict in showing
interest in both. The existence of a gray area between the
polar extremes, however, did not diminish the impact the
extremes had on stimulating what might lightly be called
"vigorous debate" within the discipline (Perkins, in preparation). The maturation of each paradigm by 1975 plus the
sharply disputed results of the PBWEE made the gray area
more difficult to occupy in the late 1970's. Proponents of

TPM continued to regard themselves as in agreement with IPM
except for those few species in which TPM was justified.
Proponents of IPM, however, came more to reject TPM as an
unworthy guide for further research. Entomologists sharing
the TPM paradigm came to criticize their IPM colleagues for
lack of vision (E. F. Knipling, Personal Interview, 1976;
Knipling, 1966a); entomologists of the IPM paradigm responded
that their TPM colleagues were articulating research goals
based on unsound scientific principles (Perkins, in preparation). The two schools of thought thus became rivals
speaking different languages and articulating different
visions for entomological research. Communication between
the two was frequently strained, awkward, and unproductive.
Emotions ran high, and the science of entomology could not
speak with one voice about the future.

The difficulties between the two research communities
notwithstanding, it must be noted that proponents of the two
paradigms together must be credited with moving entomology
away from research done under the chemical-control paradigm.
The USDA estimated (Hoffmann, 1970) that by the mid-1960's,
federal entomological research dollars on non-chemical
methods of control were:

* Biological control   14%
* Insect attractants   14%
* Sterile-male technique   12%
* Host plant resistance   7%
* Cultural/mechanical control   4%
* Basic biology   33%

Comparable estimates for expenditures by the land grant
universities are not available, but there is little reason
to believe that they differed markedly from the federal
pattern. Notable exceptions are that most research on the
sterile-male technique was done in federal laboratories; in
addition, only a few non-federal laboratories, especially
the University of California, had significant expenditures
for biological control.

The contemporary intellectual make up of the entomological research community has not been elucidated by surveys
to measure factors affecting choices of research. At the
very least, however, all three paradigmatic visions are
still to be found within those entomologists working on
applied problems.

* Chemical
* Integrated Pest Management
* Total Population Management

The measurement of the relative numbers of practitioners ad-
hering to the different schools of thought could provide
important information to policy makers concerned with pest
control. This point will be discussed more fully below.

Toward a Cultural Theory of Entomology

Conventional wisdom suggests that the power of scientific training is its ability to raise the researcher to an
intellectual level where he or she can discover the "truth"
about the natural world in an "objective" manner, i.e., in a
way that is not influenced by the multitude of political,
economic, social, aesthetic, and personal factors that shape
the other dimensions of human life. Mathematics has been
pointed to by many analysts as the key tool that liberates
the scientist from the constraints of culture.

The vision of science as a purely objective search for
knowledge, however, does not stand up to close scrutiny.
Refined laws and theories may be buttressed by an impressive
array of data and analysis that compel widespread acceptance,
even from researchers working in different cultures, locations and times. The earlier stages of scientific and
technical development, however, are significantly influenced
by cultural considerations. These factors affect the ways
technically trained people (a) formulate their research
questions, (b) gather resources, (c) elect to accept certain
data and theories and reject others, and (d) interpret their
results.

Another way of describing the impingement of external
factors on science and technology is that the boundaries of
a discipline are not rigidly defined barriers behind which
researchers retreat to pursue their work. The larger culture sends a variety of cues into the domain of scientific
and technological research. The practitioners may receive a
mixed and confused message from their fellow citizens, but
they are no more able to ignore those cues than they are to
ignore the civil and criminal laws in their non-professional
activities.

Agricultural entomology is, of course, by definition a
field of study that includes factors beyond the biology of
insects. Entomologists have been acutely aware from the
first days of their professionalization that their expertise
would be judged "true" only if it was successful in its
political and socioeconomic dimensions as well as in
biology. It is thus no major revelation to note that at
least some factors external to biology affect entomology.
The questions of importance were: Precisely what external

factors were included? Who decided? By what criteria?
Based on what assumptions? For whose benefit? The cultural
theory outlined here provides a framework by which these
questions can be answered.*

Metaphysics and Values

Everyone, including entomologists, has metaphysical
assumptions and values that impinge upon and shape their
work. To assert that metaphysics and values played a role
in entomology is not to condemn it as unscientific or non-
empirical. The training of most scientists, however, leads
them to reject most metaphysical questions as mere contem-
plation or speculation. Questions not amenable to experiments, and judgments not based on empirical observations are
held to be irrelevant to the scientific enterprise. Scientists trained in the empiricist tradition, therefore, are
frequently not sensitive to the role of the subjective in
scientific creativity.

Metaphysics and values play an important role in helping
a scientist gather and interpret data. Only a rare worker
would claim to operate purely on the basis of Baconian,
inductive methods. Instead, scientific workers use models
of the universe, sometimes implicit and unexamined models, to
construct a research plan and interpret the results. Kuhn
refers to these models as part of the "disciplinary matrix"
and notes that they can be held by the practicing scientist
with various degrees of loyalty ranging from use as a heur-
istic device to a strongly held metaphysical commitment
(Kuhn, 1977, pp. 297-298). Metaphysical assumptions, therefore, are integral and important components of scientific
research activity. Far from being perjorative to say that a
scientist is "metaphysical" in some part of his work, it is
essential that the scientist have some metaphysical assumptions in order to get anything done at all. Paul Feyerabend
(1963) argues that the absence of explicit metaphysical

---

* Thomas S. Kuhn (1977) and Harold I. Brown (1977) have been
most influential in helping me to articulate part of the
cultural theory outlined here. See especially Kuhn,
"Objectivity, Value Judgment, and Theory Choice" in The
Essential Tension, pp. 320-339; and Brown, Perception,
Theory, and Commitment: The New Philosophy of Science,
pp. 101, 105, 108-109, 166-167, 180. Also important were a
variety of studies by what are known as "externalist"
historians of science. Roy Macleod (1977) provides a useful
overview of this field.

questions in a science indicates it is headed toward a sterile, de facto, dogmatic metaphysics.

Precisely what are metaphysical concerns? Metaphysics deals with what is beyond the physical or experimental. They consist of those assumptions and beliefs about the reality of the universe (ontology) that cannot be directly tested and about fundamental causes and processes (cosmology). Some examples should help clarify their nature.

*Isaac Newton held metaphysical assumptions about time and space as independent invariants and constructed his mechanics accordingly; Einstein argued the two were related and established a new mechanics (Gillispie, 1960, pp. 141-142).

*Charles Darwin believed in a material universe governed by natural laws. The evolution of species by natural selection was an acceptable interpretation of the data of natural history for him, but alternative, creationist explanations were necessary to such scientists as Louis Agassiz who held to the contemporary Christian metaphysics (Mason, 1962, p. 424).

*J. Robert Mayer articulated in the 1840's a portion of the doctrine of the conservation of energy based largely on the metaphysical assumption that force must be indestructible (Gillispie, 1960, p. 376).

Metaphysical assumptions have changed in response to scientific and cultural developments. They are incorporated into paradigms and serve as guidelines for designing experiments, interpreting data, and making scientific judgments.

A series of fundamental metaphysical problems were implicit in applied entomology since World War II. These questions were pervasive throughout the natural sciences, so entomology was in no way unique. The most important of them can be briefly formulated as follows:

*What is the relationship between the material world and humans?

*Are there intrinsic limits to man's ability to manipulate nature?

*If limits in man's ability exist, what are they?

Simple generalizations about the range of implicit and explicit answers from entomologists to these questions can not be made. A complex spectrum of inferred and explicit opinions can be found in the literature and from unpublished correspondence and interviews. The published record of the debates between entomologists about the philosophical problems faced by their science is devoid of serious attention to these difficult problems. Yet a careful examination of the evidence available indicates that adherents to the different paradigmatic visions of entomology gave different answers to them. In the analysis below, I will attempt to sort out the correlations and demonstrate that patterns of research over a career were in part dependent upon the researchers' sense of being and process in the material world.

Entomologists performing normal science with the chemical-control paradigm were not inclined to voice a great deal of sentiment about their attitudes toward the natural world or about the relationship of humans to it. Rather, they focused on the practicality of their mission: find the cheapest and most efficient chemical to control insects and deliver the information to those people who need to control. Implicitly, they accepted the following assumptions:

*The natural world was complex in terms of how insects cause damage, but many of those complexities could be safely ignored if effective poisons were used properly.

*Man's manipulation of nature was necessary for his own well-being. The manipulation needed included the usual agricultural practices of plowing and planting. Once effective insecticides were available, they, too, became part of the "needed" manipulations. Humans were, in short, the stewards of the natural world and both could and should do what was needed to protect their interests.

*Intrinsic limits to man's ability to manipulate nature might exist, but they were far removed from the questions of controlling insects with chemicals. Insecticides had to be used with care because they were poisons, but in using them man was not treading into a situation in which they could result in a deleterious "backfire" on man's welfare.

Entomologist Clay Lyle (Mississippi State University) provided an enthusiastic vision of the chemical future in his Presidential Address to the American Association of Economic Entomologists in 1946. He believed that the effectiveness of the new insecticides such as DDT and BHC was high and that the general public was eager to follow the lead of entomologists in attacking insect problems. "Is this not an auspicious time," he asked, "for entomologists to launch determined campaigns for the complete extermination of some of the pests which have plagued man through the ages?" He then suggested targets for eradication such as the gypsy moth, housefly, horn-fly, cattle grubs, cattle lice, screwworm fly, and Argentine ant. He closed with the exhortation, "In the words of Daniel Hudson Burnham, let us 'Make no little plans. They have no magic to stir men's blood.'" (Lyle, 1947).

Lyle's attempt to rouse the troops for a concerted chemical campaign against some insect pests was not successful for reasons too numerous to review here. Indeed, E. F. Knipling, who twenty years later spoke eloquently for eradication attempts against certain key pests, recalled that Lyle's remarks had little impact on the development of his own thoughts (E. F. Knipling, Personal Communication, 1978). The failure of Lyle's exhortations notwithstanding, it is important to note that his vision of how entomologists should spend their time and effort was a general reflection of the implicit assumptions operating within the largest segment of the entomological community.

It is difficult today to find an unabashed adherent to the chemical-control paradigm among research entomologists. Even though research papers are still published in which the research design is to find how best to use particular chemicals (for example, see Staub and Davis, 1978; Linduska, 1978; Harris, Svec, and Chapman, 1978), researchers generally acknowledge that the use of chemicals carries with it a series of associated problems. Research in entomology, however, is a different entity from the complex of techniques adopted by farmers in their fields. Entomologist Robert van den Bosch estimated in 1978 that, for example, less than 10% of the cotton acreage in California was treated in a way based on the IPM-paradigm (van den Bosch, 1978, p. 173). The demise of a research community that once stoutly defended the design of research on the basis of the chemical-control paradigm has thus not yet been reflected in a transformation of pest control practice in the farm community.

The research community that developed around the IPM-paradigm developed a set of assumptions about the natural

world and man's role within it that was different from that
implicitly held in the chemical-control paradigm. Moreover,
the theoreticians of the IPM-school were more explicit about
the nature of those assumptions.

The most important assumptions made in the IPM-school
were that (a) humans are a biological species firmly embedded
in a complex ecosystem, (b) anything they do to control
insects competing with them for resources must be based on
the presupposition of man as an ecological entity, (c) man
changes the environment with technology to meet his needs,
and (d) those technologies are subject to limitations due
to human ignorance about the complexity of the environment.
The first formal presentation of the IPM paradigm in 1959
stated these assumptions as follows:

> All organisms are subjected to the physical and
> biotic pressures of the environments in which
> they live, and these factors, together with
> the genetic make-up of the species, determine
> their abundance and existence in any given
> area . . .   Man is subjected to environmental
> pressures just as other forms of life are, and he
> competes with other organisms for food and space.

> Utilizing the traits that sharply differentiate
> him from other species, man has developed a
> technology permitting him to modify environ-
> ments to meet his needs. Over the past several
> centuries, the competition has been almost
> completely in favor of man. But . . . he
> changed the environment . . . [and] . . . a
> number of species, particularly among the
> Arthropoda, became his direct competitors .
> . . Today . . . his population continues to
> increase and his civilization to advance . . .
> [and] . . . he numbers his arthropod enemies
> in the thousands of species . . .

> In the face of this increased number of arth-
> ropod pests man has made remarkable advances
> in their control, and economic entomology has
> become a complex technical field. Of major
> importance have been new developments in
> pesticide chemistry and application.

> . . . Without question, the rapid and wide-
> spread adoption of organic insecticides brought
> incalculable benefits to mankind, but it has

now become apparent that this was not an unmixed
blessing (Stern, et al., 1959, pp. 81-85).

The IPM-paradigm was thus firmly based from the beginning in an explicit concept about the fundamental principles
of the natural world and man's role in it. As the paradigm
matured in the 1970's, some important additions were made.
First, an explicit sense that man would achieve sound and
safe pest control measures by mimicing nature was articulated. Consider the following two statements, for examples:

> . . . biological control, together with plant
> resistance, forms nature's principal means of
> keeping phytophagous insects within bounds in
> environments otherwise favorable to them. They
> are the core around which pest control in crops
> and forests should be built. Biological control
> in practice . . . is . . . often possible only
> within the framework of integrated control, which
> itself usually depends upon a core of biological
> control and plant resistance (Wilson and
> Huffaker, 1976, p. 4). (My emphasis).

> Scientific pest control has always required a
> knowledge of ecological principles, the bio-
> logical intricacies of each pest, and the
> natural factors that tend to regulate their
> numbers. Today, it is more necessary than ever
> before to take a broad ecological overview con-
> cerning these problems, . . . We cannot afford
> any longer to disregard the considerable capa-
> bilities of pest organisms for countering control
> efforts . . . It is for this prudent reason that
> we must understand Nature's methods of regulating
> populations and maximize their application.
> (Smith, Apple, and Bottrell, 1976, p. 12).
> (My emphasis).

The second addition of note was hinted at in the second
quotation above: "We cannot afford any longer to disregard
the considerable capabilities of pest organisms for counter-
ing control efforts," is suggestive that man's technological
powers may be limited by intrinsic biological factors.
There is a reluctance among scientists in general and applied
scientists in particular ever to concede the existence of
intrinsic limits to man's knowledge and power. The late
Robert van den Bosch, one of the foremost theoreticians of
the IPM-paradigm, moved to such a concession in 1978 in
his criticism of chemical control:

Our problem is that we are too smart for our own good, and for that matter, the good of the biosphere. The basic problem is that our brain enables us to evaluate, plan, and execute. Thus, while all other creatures are programmed by nature and subject to her whims, we have our own gray computer to motivate, for good or evil, our chemical engine. Indeed, matters have progressed to the point where we attempt to operate independently of nature, challenging her dominance of the biosphere. This is a game we simply cannot win, and in trying we have set in train a series of events that have brought increasing chaos to the planet (van den Bosch, 1978, p. 12).

It is important to note that those entomologists who had doubts about the ability of man to manipulate the natural world at will based their pessimism on a recital of all the ills that pest control based on insecticides had demonstrated: resistance, resurgence, secondary-pest outbreaks, environmental damage, and health hazards. To these observations they added their convictions about the complexity of ecosystems and the evolutionary successes of the Arthropods over the past 300 million years. In their own literature, they seldom resorted to explicit philosophical considerations about the nature of the man-environment relationship. Rather, they presented their conclusion that man was subject to domination as one derived from an objective consideration of empirical facts. I submit, however, that such images are really assumptions that are metaphysical in nature and not subject to empirical proof. I share this assumption with the members of the IPM-school, but that in no way diminishes the importance of recognizing the presupposition for what it is. More importantly, not all entomologists who were just as upset about the problems associated with insecticides shared the presupposition about man's relationship to nature, and the type of research they pursued was markedly different as a result.

As noted earlier, Edward Fred Knipling of the USDA was (and is) the major theoretician of the paradigm I have called Total Population Management (TPM). An examination of Knipling's works over the period 1955 to the present indicates that he, too, operated on the basis of a series of assumptions that are metaphysical in nature. He shared many assumptions with his colleagues in the IPM-school: (a) humans are a biological species firmly embedded in an ecosystem, (b) anything they do to control insects competing with them must be based on the realization that man is an

ecological entity, (c) man changes the environment with
technology to meet his needs, and (d) sound pest control
will come from mimicing natural processes. The overlap of
the assumptions of the two paradigms is part of the basis
for my earlier assertion that they have much in common.
Knipling did not accept, however, the notion that technological advances were subject to intrinsic limitations. He
readily agreed that ignorance of the complexity of ecological
systems was a cause of the failure of some pest control
practices, particularly those based on insecticides, but he
was a profound optimist who believed that hard work and
dedication could solve exceedingly difficult problems in
mastering natural processes.

Knipling argued in 1965 that eradication of certain key
pests was a legitimate goal for entomological research. His
dramatic sense of optimism was shown in his conclusion:

> The development of procedures for achieving and
> maintaining complete control of specific insect
> populations will not be easy. A satisfactory
> solution to each major insect problem will re-
> quire imagination and the best scientific talent
> that we can muster. Research costs will be
> high . . . The high cost of control, the high
> losses in spite of control efforts, and the un-
> desirable side effects of current methods of
> control obligate us to take an entirely new
> look at some of the most costly and most trouble-
> some of our insect problems. There is ample
> justification for taking bold and positive
> steps in our research efforts . . . These are
> the reasons for my interest, my confidence, and
> my enthusiasm . . . (Knipling, 1966b).

In 1978, Knipling again reiterated his supreme confidence in the prospects for successful research. He outlined
three levels of control: (a) eradication when technically
feasible and economically justified; (b) area-wide or
ecosystem-wide management of some major pests; and (c)
critical monitoring of pest populations and application of
control measures when needed. The three strategies are
listed in decreasing order of difficulty, and Knipling
acknowledged that the eradication notion was ". . .
probably the most controversial among members of the
entomological community." Nevertheless, he argued that continual improvements in technology required a continual re-
appraisal of the technical feasibility of eradication
efforts. His sense of optimism, indeed his faith in the

forthcoming fruits of technological innovation were again
articulated:

> I have a great confidence in the ingenuity of
> our young scientists to perfect the technology
> necessary to put sound principles of insect
> suppression into practice in future years.
>
> . . . I see real opportunities for relegating
> many of the more persistent and costly pests to
> a status of minor importance economically, and
> in an ecologically sound manner, by reducing
> total populations on an ecosystem basis in an
> organized and coordinated way, using some of
> the approaches and principles of suppression
> discussed.(Knipling, 1978).

Knipling's confidence that technology could be developed
to the point of totally managing an insect pest, even to the
point of eradication, must not be interpreted to mean that
the adherents of the IPM-paradigm were mere pessimists who
doubted the ultimate successes of their own creative research efforts. Far from it, they were just as confident
of their chances of success as adherents of the TPM-paradigm
were.

The differences between the two schools of thought rest
on more subtle points: Most adherents of the IPM-paradigm:
(a) saw no particular need to reduce a pest population to
zero, (b) viewed eradication efforts as diversionary from
better avenues of research, and (c) believed eradication
efforts would almost invariably prove unworkable, especially
for well-established and widely-distributed insects. The
IPM-school was content, in other words, to suppress a pest
species below economically damaging numbers and then do no
more than necessary to keep it there.

The issue of eradication, therefore is the heart of the
difference between the two schools. Eradication is the
ultimate in ecosystem management in that once a species is
removed from an area, the ecosystem is qualitatively
different in perpetuity. The reduction in numbers of a pest
species resulting from manipulations derived from the IPM-
school also changes the ecosystem, but the continued
presence of the animal in the area means the change is
reversible. The high reproductive capacities of insects
would cause the pest to regain high population densities if
suppression techniques were removed.

The change in the ecosystem from eradication has profound implications for human behavior in that fewer constraints remain on human activities. Specifically, a farmer
who is freed from ever having to worry about a pest can
alter his production practices without having to consider
the implications of the change for its effect on the former
pest insect. A pest control scheme in which eradication is
never attempted or achieved is destined to be needed in
perpetuity because the insects will always be a potential
problem. A farmer thus has no hope of ever being freed from
the constraints imposed by the presence of the insect.
Knipling was highly conscious of this limitation of the IPM-
paradigm and was unwilling to accept it:

> [Carl B. Huffaker and Ray F. Smith, University
> of California] are not thinking integrated control
> in the sense that I am. I'm thinking integrated
> control in the sense that you're taking advantage
> of the characteristics of different systems and
> putting them together for total management of a
> population. They're looking at integrated
> control . . . [as being] based on assessment of
> economic threshold levels and not to use control
> measures until they reach that goal . . . Now,
> I maintain that we'll never solve some of these
> insect problems that way, (Knipling, Personal
> Interview, 1976). (My emphasis)

A second type of difference between the adherents of the
two schools centers on the problems of the legal and moral
rights of other species. During the 1950's and 1960's,
little mention of rights of other species could be found anywhere in the literature of the industrialized world and
certainly not in the writings of applied entomologists. The
emergence of the environmental movement in the late 1960's,
however, brought the notion of such "rights" into the arena
in which debates on insect control were fought (see, for
example, Murphy, 1971). Eradication came to be seen by a
few entomologists as a concept posing serious questions for
their discipline in terms of the rights of other species.
The recent advent of such questions makes it impossible to
do more than briefly summarize the current hazy state of the
debate.

Proponents of eradication (TPM) implicitly assumed that
the target of annihilation had no rights in the treatment
area; since these entomologists never seriously considered
the global eradication of an insect, there was some ambiguity
surrounding their implicit assumptions of rights outside the
targeted eradication zone. Knipling believed eradication of

a native species might be ecologically damaging, but his concern was for deleterious consequences for the ecosystem, not
the target insect (Knipling, 1978, p. 50). Proponents of
IPM, on the other hand, had mixed reactions about the rights
of other organisms. Dale Newsom (Louisiana State University)
believed no moral principle was involved; he would, for
example, be glad to eradicate the boll weevil, but he doubted
the effectiveness of the proposed technology (Newsom, 1978).
Paul DeBach, one of the foremost advocates of biological
control, was not opposed to eradication on moral grounds because extinction is a natural process. He like Newsom,
raised questions of practicality and the effect of eradication on the ecosystem as a whole (DeBach, 1964). Robert
Rabb (1978) moved closer to a principled objection to
eradication: "The use of the [technological] power is a
tremendous responsibility and must be done without arrogance
and with a subtle sensitivity, if not a reverence, for the
value of all life." Entomologist Robert L. Metcalf of
Illinois occupied the polar position with an explicit, metaphysical assertion, ". . . I do firmly believe that species
should be regarded as sacred and man indeed has no right or
reason to destroy them" (Metcalf, Personal Communication,
1978).

The above discussion on the metaphysical assumptions and
presuppositions contained within contemporary efforts to
innovate in entomology began with the assertion that two
related but in some ways rival paradigms were developed
during the late 1950's and 1960's. If the philosophical discussion just presented is accepted, then clearly one source
of differences between adherents of the two paradigms is
philosophical in nature. Succeeding sections will raise the
possibility that other differences exist. Furthermore, I
have presented no argument about the possible sources of
the philosophical differences; such discussion is beyond the
scope of this paper and will be developed elsewhere. The
importance of metaphysical presuppositions in entomology
appears so strong, however, that it is worth venturing some
labels in order to facilitate discussion about the issues
raised. Labels can both obscure and illuminate, so their
use is not an unmixed blessing. Nevertheless, I will propose
some with the hope they will help, not hinder, further
thinking.

Both the IPM and TPM paradigms are embedded in a matrix
of naturalism. Both see man as an element of the natural
world and both articulate their visions in terms of ecosystems and learning how to mimic nature in controlling
insects. The crucial differences between the two lie in the
position accorded man: The IPM-paradigm stops short of

venturing for total mastery of nature as epitomized in the
notion of eradication. The TPM-paradigm makes that step
beyond IPM and argues that total mastery of ecosystems, up to
and including qualitative adjustments of the species composition, is the vision towards which entomologists should bend
their efforts. The crucial difference between them thus is
the position of man within the biosphere: He is not the
total master in IPM; he dares to be so in TPM. It is for
this reason that I propose "naturalistic" as a name for the
underlying presuppositions of IPM and "humanistic" for TPM.
The meanings of each term are as follows:

Naturalistic: A belief system that man is a part of
the biosphere but that he cannot be
the total master of it. He may manipulate for his own benefit, but there
are intrinsic limits to his manipulative powers that reside in the
properties of the material world.

Humanistic: A belief system that man is part of
the biosphere and that he can be master
of it. He may manipulate it for
his benefit, and there are no intrinsic limits to his manipulative
powers that reside in the properties
of the material world. The limits
such as they are derived from his
current ignorance of natural
processes.

The foregoing discussion aimed to establish the fact
that different philosophical assumptions have been present
in agricultural entomology. The real importance of such
assumptions, however, lies in their role as components of
paradigms. Our effort to understand the importance of
assumptions in the entomological research community thus
leads us to ask what effect did the different paradigms
have on individual research careers? More specifically, do
individuals working from different paradigms exhibit differences in (a) the types of data they choose to gather, (b)
how they interpret that data, and (c) the types of field
practices they advocate?

The pattern of research problems to which individuals
dedicated their careers provides some insights into the
guiding role paradigms played. The interpretation of such
patterns is difficult, but the careers of Carl Barton
Huffaker and Edward Fred Knipling can each be seen as

reflections of the differing paradigms they helped develop
and then used.*

Huffaker obtained his Ph.D. in 1942 from Ohio State
University and took his first full-time position with the
University of Delaware on problems in mosquito control. His
approach was primarily ecological, and he continued the work
from 1943 through 1945 with the Institute of Inter-American
Affairs in Latin America. The difficulties of living in the
tropics plus the dissatisfaction of working with engineers
and medical doctors, however, led him to seek a new position.
He obtained a job under Harry Scott Smith at the University
of California, Berkeley, with an assignment of cooperating
with the USDA on an attempt to control the Klamath weed with
insects.

Huffaker and James K. Holloway of USDA were dramatically
successful against the Klamath weed, and Huffaker still
remembers the elation as one of the two most exciting pieces
of research in his career. His interests in the ecology of
plants and insects plus this success encouraged him to de-
vote his research activities to biological control, population ecology, and, later, integrated pest management. Of
the 165 papers he wrote between 1941 and 1978, 157 are
clearly devoted to this perspective. His work formed part
of the base on which the IPM-paradigm was based, and he was
one of the theoreticians who helped articulate that paradigm.
The remaining included three oriented to chemical control
and four that fall outside the categories selected here as
relevant. His service as Director of the "Huffaker Project"
from 1972-1977 was a capstone to many years of work attempting to forge a way out of the dilemmas associated with
insecticides.

Knipling's work patterns were more complex. He joined
the USDA staff in 1931 after completing his master's degree
at Iowa State University. He obtained his Ph.D. after World
War II from the same University. His first assignment was
on surveys of screwworm fly populations in Texas. He began
publishing in 1934, and established a record of contributions on insects parasitic on livestock by the end of the
decade. He also gave considerable evidence of research
leadership abilities and moved up the hierarchy in USDA.

---

* The descriptions of the careers of Huffaker and Knipling
are derived from personal interviews, their bibliographies,
and their curricula vitae. In addition, each gave helpful
comments on a draft version of this analysis. Neither
should be held responsible for this analysis, however.

In 1942, he was named Director of the Orlando, Florida, laboratory designated by USDA as the locus for work to protect American and Allied troops from insects. This laboratory was the first to test DDT in the U. S. (Perkins, 1978b). Knipling was therefore one of the first U.S. entomologists to observe the dramatic killing powers of the new insecticides. Knipling's research record from 1942 to 1950 reflected the problems and promises of DDT and other new insecticides that followed it. His unpublished correspondence and recollections, however, indicate that between 1937 and 1950, he was thinking of a revolutionary insect control idea, the sterile-male technique. Knipling conceived the idea in the late 1930's, but the lack of sterilizing methods plus the war prevented him from working on it. His outstanding achievements at Orlando, however, resulted in his appointment as Chief of the Division of Insects Affecting Man and Animals in Washington. Administrative responsibilities made it possible for him to pursue the funds necessary. He enabled his colleague Raymond C. Bushland to begin experiments on sterilizing with x-rays in 1950. The first publications on the use of sterile-males in field eradication trials came in 1955 when Knipling and numerous colleagues announced that the sterile-male technique had rid the island of Curacao of screwworm flies. Even more dramatic success followed in 1958-1959 when the USDA and the Florida livestock industry used the sterile-male technique to rid Florida and the southeastern states of screwworms. The method demonstrated its ability to suppress the total population of screwworms over the vast grazing lands of Texas and northern Mexico starting in 1962 (Scruggs, 1975; Perkins, 1978a).

Knipling's elation at the success of the sterile-male techniques combined with his sense of general progress in other entomological areas caused him to turn toward the development of comprehensive strategies in which total populations of certain key pests would be attacked on a coordinated basis with the goal of at least markedly suppressing if not actually eradicating the offending creatures. Approximately one-half of the nearly 100 papers he published between 1955 and 1976 dealt explicitly with this theme. The remainder reflected his general administrative responsibilities.

Despite some superficial similarities, the differences between Knipling and Huffaker's career research patterns are real and of fundamental importance. The fundamental similarity was that both advocated a reduced reliance on insecticides and the adoption of multiple control techniques in a coordinated package. The fundamental difference,

discussed earlier, was that Knipling believed eradication
for some pests was a legitimate goal to entertain on
technical, environmental, and economic grounds. The mastery
of the natural environment implicit in eradication reflected
Knipling's humanistic concept of man's place in nature.
Other contrasts to Huffaker's work were consistent with
this fundamental distinction. Knipling was not content to
rely on natural control agents alone because he believed
that many of the nation's pests do not yield to them, particularly in ecologically disrupted agroecosystems. He
proposed the mass rearing and release of parasitic and
predatory insects and pathogens as one approach to pest
management that would make the use of natural enemies more
effective and dependable. In addition, he continually
explored the potential of the sterile-male and other genetic
techniques, which he felt had not been fully realized. The
theme which united his work was the assumption that
entomologists should take an active role in supplementing
natural control in order to better serve human interests.

Huffaker, in contrast, preferred research on strategies
that relied on natural controls operating with minimal
human intervention. His major goal early in his career was
to find and introduce parasites, predators, and pathogens
that sustained themselves and provided adequate suppression
of the pest species. Later, he turned more to the question
of integrating biological control with other techniques of
suppression. He was a skeptic that technology could master
natural process to the extent that well-established pests
could be eradicated. Instead, he preferred naturalistic
schemes of pest suppression in which human interests could
be served without a total mastery of the natural environment. He was not opposed in principle to eradication, but
he did not think it necessary for human welfare.

Knipling and Huffaker were only two of many prominent
leaders in post-World War II entomology, but they can be
thought of as representatives of different schools of
entomological research. They had leadership roles, which
implies the existence of "followers." Why would a researcher choose to join one school in preference to another?
Thomas Kuhn provided a most provocative statement that can
serve as a working hypothesis: ". . . scientists who share
the concerns and sensibilities of the individual who dis-
covers a new theory are ipso facto likely to appear disproportionately frequently among that theory's first sup-
porters" (Kuhn, 1977, p. 328). (My emphasis) Kuhn is suggesting that the followers were predisposed by "concerns
and sensibilities," or what I would call metaphysical
assumptions, to align themselves with a school. The leaders

were those whose vision of the natural world was most well developed and who could articulate a comprehensive program of research that made sense to the followers. It is in this light that we see the importance, indeed the indispensability, of paradigms and metaphysics to scientific activity.

The fact that there were opposing paradigms in entomology probably enriched the field because of the competition engendered between them. At the same time, however, the clash of opposing philosophies spilled into the policy arena and resulted in confusion there. The question was and still is how to find the way out of the chemical-control paradigm. Were the IPM and TPM paradigms (a) mutually compatible, (b) separate but equal, or (c) separate but unequal? We will return to these points below.

The Clients: Whom Does Entomology Serve?

Applied agricultural entomology is mission oriented and must solve practical problems for clients to justify its existence. Who were the clients? It is too simplistic to assert that "the general public" constituted the clientele. Despite continued celebration of America as the great melting pot, we are still a society divided by class, race, religion, sex, national origin, and geography. Recent studies by Hightower (1973), Perelman (1977), Noble (1977), van den Bosch (1978) and others have amply destroyed any mythical vision that the innovators of technology anywhere have served the "general public." Rather, usually innovation directly serves special interests first and only indirectly does it perhaps begin to aid others. Innovation may serve the "public interest," but in some cases, it harms some groups by, for example, displacing them from employment. The problem, then, is to untangle just who entomology served. Once the clientele is identified, we can begin to ask how the nature of the client's interests affected the science and technology of insect control.

The vast majority (over 80% in 1973) of entomologists were salaried professionals working in non-profit govern-ment or university organizations. The only significant number (12%) of entomologists working in the private, profit-motivated sector were those with the chemical industry (Hardee and Tomita, 1973). They devoted most if not all of their time to the search for new insecticides and the main-tenance of existing ones. The entomologists responsible for generating the knowledge on which less-chemically oriented control technologies could be based came therefore almost entirely from the public sector, especially USDA and the land grant universities (the USDA/LGU complex).

The location of most entomologists in the public sector created problems for entomologists. Their research laboratories were responsible, variously, to the Congress, the President, Boards of Trustees, State Legislatures, the Governors, and, ultimately, to the taxpayers and voters. The laws under which the institutions were established, however, explicitly directed them to study preferentially the problems facing "farmers."* Unfortunately, even the term "farmer" is not particularly enlightening, because farmers themselves were heterogeneous, especially with regard to wealth, race, and geographic location. Both Hightower (1973) and van den Bosch (1978) have argued persuasively that in general the USDA/LGU complex has served primarily those farmers who were white and above average in wealth and thus reinforced the trend to fewer people engaged in agriculture and larger average sizes for farm units.

Entomologists within the USDA/LGU complex responded to the needs of their clients by orienting their research to serve farmers' short-term needs for profits. In the period of Euphoria and the Crisis of Residues (1945-1955), most researchers turned to insecticides because in the short-run they believed insecticides served profits better than other methods. Experiments leading to chemical control techniques frequently did not measure profits directly. Instead, the entomologists made the assumption that high kill rates of insects gave higher yields, which meant higher grower profits. As a first approximation, the assumption had merit, but more refined work sometimes indicated the assumption was not always true. For example, Paul DeBach (1951) saw quickly that the new insecticides such as DDT could destroy highly profitable biological control schemes in citrus groves. Huffaker realized similarly that there was no profitable alternative to Klamath weed control other than to find some biological method of controlling it.

In a similar vein, during the period of Confusion and the Crisis of the Environment, entomologists turned away from chemicals as the only or major component of insect

---

* The Morrill Act establishing the land-grant colleges was enacted in 1861. The Hatch Act establishing the agricultural experiment stations passed in 1887. In 1914, the Smith-Lever Act established the extension service. The U. S. Department of Agriculture started as an adjunct of the Office of the Patent Commissioner in 1836 but was upgraded to cabinet level in 1889. The Congress intended for each of these institutions to serve and promote commercial farming with many programs including research and education.

control technology because the chemicals had shown their inability to protect profits. E. F. Knipling recognized
early the significance of resistance as a threat to farmer
profits. He exercised genuine leadership in moving the USDA
laboratories into alternative lines of inquiry. It is
important to note that those who moved away from research on
insecticides in the late 1950's did not do so initially because of complaints about environmental hazards. Entomologists as a class were slow to accept the fact that the chemicals can cause trouble even when used as intended. Rather,
it was the spectre of technological failure measured in lost
profits that motivated them.

The long legal and social tradition that the public
sector provided entomological research for the private
sector had in it a contradiction that ultimately created a
crisis for the profession. The taxpayer/voter paid the bill
for entomological research and thereby gained a legitimate
interest in the nature of that research and the types of
technologies created from it. As long as the interests
of farmers and non-farmers were synonymous, there was no
conflict.   The increasing trend for chemical control to
dominate insect-control technology led to a split in the
perceived interests of the two groups. Farmers continued
to place short-term profits above all other concerns, and
insecticides provided the best tool to meet their objectives
so long as resistance and other problems did not interfere.
Consumers, however, began to perceive that their needs for
inexpensive food and fiber were balanced by a need not to be
poisoned.

The split between farmers and consumers began early in
the twentieth century in the case of apples, pears, and
other fruits and vegetables. Farmers fearing loss of profits argued for higher tolerance for lead and arsenic, and
consumer and health advocates argued for less (Whorton,
1974). Entomologists lined up with the farmers because that
was the group they perceived as their clients and patrons.
The intensity of the struggle over residues increased after
World War II, and the scope of the battle was significantly
enlarged when Rachel Carson dropped her bombshell of Silent
Spring. Environmental safety became a value for consumers/
voters/taxpayers, and their claims upon entomologists became
stronger. The pressures put upon insecticides directly, and
therefore entomologists indirectly, between 1962 and 1972
came from the U.S. Department of Health, Education and

Welfare (1969); state departments of natural resources;[*] and from groups of private citizens (Dunlap, 1975). The symbiotic relationship between the profession of entomology and farmer-businessmen that had developed over 70-100 years was suddenly threatened.

A partial destruction of the relationship occurred in the first three years of the 1970's. Registration of pesticides was removed from the USDA, the home in Washington for farmer interests, and placed in the newly created Environmental Protection Agency. In addition, the new Federal Environmental Pesticide Control Act of 1972 enfranchised non-target organisms including people with stronger rights to protection against the unintended effects of pesticidal chemicals. The reorganization and new legislation together formalized the claims of non-farmers that they were entitled to at least some voice in judging the acceptability of expert entomological knowledge. No longer were the short-term interests of farmers the only determinant in shaping entomological research and practice.

The non-farming sector has made only slight progress in exercising its rights. Their support for Carl Huffaker's integrated pest management (IPM) research was crucial (see Nixon, 1972, p. 7). Even in the case of IPM, however, the interests of farmers came first. Entomologists who were in the vanguard of promoting IPM carefully couched their language to emphasize how the profits of the farmer must be the primary objective (Huffaker, 1971). They believed to do otherwise would be useless because farmers would not voluntarily adopt something less profitable than their current technologies. Allusion to environmental values was significant, but the IPM movement never challenged the legitimacy of the farmer's right to place a high premium on short-term returns. In addition, the evidence suggests that the only instances in which farmers have adopted IPM are those in which their short-term profits were either increased, chemical alternatives were too expensive, or in which chemicals were so threatened by resistance, secondary-pest outbreaks and residue problems that IPM provided the only way out of the dilemma (Huffaker and Croft, 1978). The IPM movement has not yet attracted significant interest from persons outside the entomological profession, congressmen from rural areas, and members of the farming industry, especially cotton. It, too, however, is predicated on the

---

[*] Rachel Carson relied on reports of wildlife damage prepared by, for example, the Southeastern Association of Game and Fish Commissioners (Carson, 1962, p. 327).

belief that industry profits will be well served by successes
of its efforts (Knipling, 1978).

Entomology the Profession: Where are the Boundaries?

The ambiguity about the identification of the clientele
of entomological expertise was correlated with a parallel
confusion over the boundaries of entomology as a field of
research and study. Entomologists included questions of
hazards from residues and damage to the environment in the
field of entomology, but their close identification with the
interests of farmers made it difficult for them to pursue
these issues with vigor or enthusiasm. The result was that
challenges to entomology from outsiders left the profession
continually on the defensive. The following examples show
how entomologists were unable to gain an upper hand in
matters:

> *1950-1952: Entomologists argued against the
> need for new legislation to protect consumers
> from residues; Congress disagreed and amended
> the Federal Food, Drug and Cosmetic Act in
> 1954.

> *1962-1963: Entomologists responded to Silent
> Spring by acknowledging much of Carson's argu-
> ment and stating they were already taking care
> of all problems; the President's Science
> Advisory Committee argued for stronger actions
> on the part of USDA and other federal agencies.

> *1972: Entomologists argued against deregis-
> tration of DDT for its most prominent agricultural
> use, cotton; the Environmental Protection Agency
> disagreed and deregistered DDT on cotton and most
> other commodities.

The end result was that the questions that entomologists
must consider when developing and evaluating insect control
technologies were enlarged to include more conscious and
explicit awareness of the environment. Each addition to the
field of inquiry reflected the interests of non-farming
clients of entomological knowledge. Those few entomologists,
such as Paul DeBach and the late Robert van den Bosch, who
actively and publicly supported the inclusion of a strong
input from environmental activists into entomology were

exceptions to the norm and dramatized the strength of professional orientation to farmer interests.*

The forced addition of subject matter to entomology was a source of frustration to the professionals. Following 1955, they had less assurance that they knew <u>what</u> questions were worth asking, and they had conflicting criteria about how to defend the <u>validity</u> of the answers. Both their alternatives were unattractive: Either they could continue the old ways in which they concentrated almost exclusively upon the interests of farmer clients, or they could alter their research patterns to account for the new interests and demands of non-farmer clients. The former path was politically hazardous because of the continued pressure of non-farm constituencies. The latter risked alienating the clients who had been their source of support for so long and who would be the ones to adopt or reject the new control methods developed. Furthermore, branching into new questions required dealing with areas foreign to their training and philosophical traditions. The crisis was philosophical in nature, but political and sociological in its impact on the professionals.

The questions about health hazards and environmental damage brought entomologists face to face with a matter of utmost importance: who is a peer and therefore entitled to make judgments about entomology? Who, in other words, is qualified to speak on matters epistemological? Many people not trained in entomology made a great deal of noise about the profession, especially its problems, but no profession can survive as a body holding special knowledge and privileges unless it can control the certification rites to that position.

It is still too early to tell what the resolution to the crisis is. Only a few features are evident at this point: First, farmers still have a privileged status among the non-peers. Some environmental activists, such as the Environmental Defense Fund, have joined in active support of entomologists in the IPM-paradigm. Second, the Entomological Society of America organized in 1970 the American Registry of Certified (later Professional) Entomologists (ARPE), a group with its own by-laws, requirements for membership, and code of ethics. ARPE is striving to be the locus of certification for the profession in matters concerning control operations,

---

* De Bach and van den Bosch, for example, have been two of the few professional, applied entomologists willing to testify for the Environmental Defense Fund.

but ARPE has not yet attracted all applied entomologists. At the present time, it has no hold on licensing practices comparable to the American Medical Association or the American Bar Association.

The quest for professional legitimacy symbolized by ARPE has roots going back to the emergence of the profession in the late nineteenth century. The issues today are similar to those of years ago: Eliminate quackery; establish the need for the profession in the society at large; and control the boundaries of and certification to the discipline. It is possible that one outcome of the struggle for professional autonomy will be the establishment of a new discipline, pest management specialist. Several universities have curricula and degree programs for the activity, and there is considerable informal discussion of the subject among the entomologists, specialists in weed control, plant pathologists, and nematologists (Apple and Smith, 1976, pp. 186-187).

Implications for Public Policy

The policy problems of today have evolved simultaneously with the developments in entomological science described above. Chemical control techniques continue to dominate in agricultural practice, but many research entomologists have abandoned the chemical-control paradigm as a guide to their activities in the field and laboratory. There is a consensus that it would be desirable to move insect control practices by farmers into a variety of non-chemical activities, and the research efforts of the public-sector entomologists are seen as one of the keys to the success of such an effort. A recent study (NAS, 1975a) outlined the major policy problems involved in the effort to transform entomological technology at the farm level. The historical review presented here provides some additional insights into the fundamental question of what we can expect from entomological research activities as they are currently developing. The points briefly outlined here have not been seriously dealt with in other policy studies. The cultural theory outlined above indicates to this author that the problems facing entomology cannot be resolved without consideration of them.

The first point is that the policy makers have not recognized that two traditions have emerged since the mid-1950's as guides for research: the IPM- and TPM-paradigms. Members of the two schools of thought have engaged in a certain amount of bickering within the profession and in the policy arena, certainly nothing unusual or unhealthy in any scientific field. In fact, entomology is possibly better off

for the argumentation. Policy makers, however, have a
different set of concerns, namely, how should limited research funds be allocated? Specifically, should differential
allocations be made to experiments designed in the different
paradigms? This fundamental policy question really can't be
answered yet because we still have an incomplete sense of
the relationship between the two paradigms. Are they compatible with IPM being the fall-back position from TPM?
Adherents of the TPM-school maintain that efforts to manage
and perhaps eradicate certain key pests on an ecosystem-wide
basis will produce information useful to IPM if the more
ambitious goal is not achieved. Some adherents of the IPM-
school have suggested the two paradigms may be incompatible
(Rabb, 1978). The question of compatibility is not trivial
and requires further analysis of the sociological structure
of the entomological research community and of the philosophical assumptions underlying the two paradigms. There
is the further question of the relationship of each of the
two paradigms to the interests of the farming and non-
farming sectors of society. Are control techniques emanating
from each likely to serve social interests equally? This,
too, is a difficult question requiring further analysis.
Alternative technological scenarios need to be created and
examined for their social impacts; likewise, as Stockdale
suggests in this volume, we should have a better idea of
where we want to go and ask how each paradigm might serve
such larger social visions.

Entomologists have not seriously examined their paradigms or their techniques for philosophical implications and
values questions. This is a problem that pervades most if
not all of the applied sciences. Engineering and medical
schools have made some efforts to establish courses and
programs dealing with human values and technology, but the
agricultural schools have made only meager comparable moves.*

---

* A recent survey (AAAS, 1978, pp. 3-32, 90-98) identified
117 programs designed to explore the values questions
implicit in science and technology. Only 11 universities
with agricultural colleges (Cornell, Iowa State, Michigan
State, New Mexico State, North Dakota State, Oklahoma State,
Pennsylvania State, Purdue, Florida, Wisconsin-Madison, and
Utah State) offered 12 of 68 programs on "Science, Technology,
and Human Values." Significantly, not one of the 12 programs
was described with any specific component focused on the
agricultural sciences. Several had environmental and conservation components, but most were oriented to the science
and technology of the engineering schools. The University
of California (Berkeley) program in Conservation and Resource

It is necessary to establish such programs so that established and new entomologists can gain some appreciation for the importance of values, metaphysical assumptions, and paradigms in the conduct of scientific research. Research from such programs could help in answering the policy questions posed above.

The second issue centers on who does or ought the entomologist serve. Service dominated by the interests of large-scale commercial farming is no longer adequate. Non-farmers have a right and a need to participate in the shaping of entomological research and practice. It may be that some entomological knowledge can serve both farmers and non-farmers alike, but the assumption that the interests of the two are identical should be abandoned. The relevant committees of legislative, executive, and academic bodies should include the legitimate interests of non-farmers in planning and implementing research decisions. Such participation will be difficult to achieve but is sufficiently important to warrant serious efforts.

The third issue is the need to evaluate the nature of professional organizations serving entomologists, particularly the Entomological Society of America and the American Registry of Professional Entomologists. Entomologists have behaved similarly to other professional groups in America when their field of knowledge was challenged by outsiders: they moved to upgrade the standards of the group and establish rigorous internally-governed criteria for membership. The upgrading of standards was important, but such activity only goes part-way to solving the problem. One of the most significant sources of disenchantment with entomologists over the use of insecticides was the perception of them as a closed society that was immune from the concerns of the general public that supported it. What is also needed, then, is an effort to open the profession to more inputs from outsiders. ESA and ARPE should establish mechanisms whereby

---

Studies and Cornell University, College of Agriculture may be the only two places where departments located in an agricultural college offer courses with heavy emphasis on values questions.

Another indication of the dearth of attention given to values questions is that of 52 courses designed to cover professional ethics for new professionals, not one of them dealt with agricultural professionals. All were concerned with engineering, psychology, medicine, and the allied health sciences.

interested non-professionals can give advice on an ongoing
basis to the professional organization. In this way,
entomologists could maintain the contact with the broad
interests they need. It is also likely the professionals
would benefit because their outside advisors might become
advocates for new and innovative types of research. The
creation of a more open entomological profession would in
itself be a significant advance that could serve as a model
to other professional groups facing similar problems.

Acknowledgments

Many people have given generously of their time to help in preparation of this article. I'm particularly indebted to P. L. Adkisson, J. R. Brazzel, T. B. Davich, Paul DeBach, Charles Lincoln, L. D. Newsom, C. R. Parencia, Reese I. Sailer, and J. R. Smith, all of whom shared their first-hand experiences in entomology through long interviews and some of whom provided access to their unpublished correspondence. I am particularly indebted to Carl B. Huffaker and E. F. Knipling, both of whom consented to be interviewed, shared correspondence with me, and criticized an early version of this paper. Both helped eliminate errors. Others who provided invaluable criticisms and comments were William H. Newell, Barbara B. Perkins, and David Pimentel. Some of those interviewed for this study may disagree with my evaluation of their profession or work. They are certainly not to be held responsible for my judgments. Responsibility for errors and obfuscations likewise belong to me.

Student research assistants who labored on various parts of this project were Jeffrey C. Page, Keith M. Johnson, G. Alex Echols, David M. Soloway, Joseph Albrechta, and S. Martijn Steger. Susan L. Tiefel helped enormously by typing transcripts of the interviews with Knipling, Huffaker, and DeBach, and in other ways.

Some of the materials incorporated in this work have been developed with the financial support of the National Science Foundation (SOC 76-11288) through contract from Miami University to the University of California. However, any opinions, findings, conclusions, or recommendations expressed herein are those of the author and do not necessarily reflect the views of the University of California, Miami University, or the National Science Foundation. Financial assistance of the Josiah Macy, Jr. Foundation is also gratefully acknowledged.

Special thanks go to the Division of Biological Control, University of California, which hosted me while this article was written. I also thank Myron J. Lunine, Dean of the School of Interdisciplinary Studies, Miami University, for accommodating my leave from teaching duties while the article was prepared.

References

AAAS. 1978. EVIST Resource Directory. AAAS, Washington. 208 pp.

Agricultural Research Service. 1962. Boll Weevil Research Laboratory. USDA, Washington. 23 pp.

Apple, J. L., and R. F. Smith. 1976. Progress, problems, and prospects for integrated pest management. pp. 179–196 in Integrated Pest Management. J. L. Apple and R. F. Smith, eds. Plenum Press, New York.

Baldwin, I. L. 1962. Chemicals and pests. Science 137: 1042–1043.

Benedict, M. R. 1953. Farm Policies of the United States, 1790–1950. The Twentieth Century Fund, New York. 548 pp.

Bishopp, F. C. 1951. Testimony before U. S. Congress, House Select Committee to Investigate the Use of Chemicals in Food Products. Hearings 81: 2.

Brazzel, J. R., T. B. Davich, and L. D. Harris. 1961. A new approach to boll weevil control. J. Econ. Entomol. 54: 723–730.

Brazzel, J. R., and L. D. Newsom. 1959. Diapause in Anthonomus grandis Boh. J. Econ. Entomol. 52: 603–611.

Brooks, P. 1972. The House of Life, Rachel Carson at Work. Fawcett Publications, Inc., Greenwich. 303 pp.

Brown, A.W.A. 1961. The challenge of insecticide resistance. Bull. Entomol. Soc. Amer. 7: 6–19.

Brown, H. I. 1977. Perception, Theory, and Commitment: The New Philosophy of Science. Precedent Publishing, Inc., Chicago. 202 pp.

Brues, C. T. 1947. Changes in the insect fauna of a New England woodland following the application of DDT. Harvard Forest Papers 1: 2–18.

Carson, R. 1962. Silent Spring. Houghton Mifflin Company, The Riverside Press, Boston. 368 pp.

Chapman, P. J. 1963. The use of chemicals to control pests. New York State Hort. Soc. Proc. 108: 168–175.

Conant, R. 1944. No joy in an insect free world. Entomological News 55: 258-259.

Corbet, P. S., and R. F. Smith. 1976. Is integrated control realistic? pp. 661-682 in Theory and Practice of Biological Control. C. B. Huffaker and P. S. Messenger, eds. Academic Press, New York.

Cottam, C. 1946. DDT and its effect on fish and wildlife. J. Econ. Entomol. 39: 44-52.

Darby, W. J. 1962. Silence, Miss Carson. Chem. Eng. News 40: 60.

DeBach, P. 1951. The necessity for an ecological approach to pest control on citrus in California. J. Econ. Entomol. 44: 443-447.

DeBach, P. 1964. Some ecological aspects of insect eradication. Bull. Entomol. Soc. Amer. 10: 221-224.

Doutt, R. L., and R. F. Smith. 1971. The pesticide syndrome—Diagnosis and suggested prophylaxis. pp. 3-15 in Biological Control. C. B. Huffaker, ed. Plenum Press, New York.

Dunlap, T. R. 1975. Scientists, Citizens, and Public Policy. Ph.D. Thesis, University of Wisconsin. 370 pp.

Dunlap, T. R. 1978. The triumph of chemical pesticides in insect control, 1890-1920. Environ. Review 5/78: 38-47.

Dutky, S. R. 1952. The milky diseases. pp. 75-115 in Insects, The Yearbook of Agriculture 1952. Government Printing Office, Washington.

Entomological Society of America Review Committee. 1973. The pilot boll weevil eradication experiment. Bull. Entomol. Soc. Amer. 19: 218-221.

Federal Register. 1949. September 17, p. 5724.

Feyerabend, P. 1963. How to be a good empiricist—A plea for tolerance in matters epistemological. pp. 3-39 in Philosophy of Science, The Delaware Seminar, Vol. 2, 1962-1963. Interscience Publishers, New York.

Flint, M. L., and R. van den Bosch. 1977. A Source Book on Integrated Pest Management. Department of Health, Education, and Welfare, Washington. 392 pp.

Forbush, E. H., and C. H. Fernald. 1896. The Gypsy Moth. Wright and Potter Printing Co., Boston. 495 pp.

Gillispie, C. C. 1960. The Edge of Objectivity. Princeton University Press, Princeton. 562 pp.

Hall, I. M. 1963. Microbial control. pp. 477-517, Vol. 2, in Insect Pathology, An Advanced Treatise. E. A. Steinhaus, ed. Academic Press, New York.

Hardee, B., and K. Tomita. 1973. A Survey of Scientific and Professional Characteristics of the General Membership of the Entomological Society of America. Center for Research in Scientific Communications, Johns Hopkins University, Report #25. 26 pp.

Harris, C. R., H. J. Svec, and R. A. Chapman. 1978. Potential of pyrethroid insecticides for cutworm control. J. Econ. Entomol. 71: 692-696.

Hawley, I. M. 1952. Milky diseases of beetles. pp. 394-401 in Insects, The Yearbook of Agriculture 1952. Government Printing Office, Washington.

Heady, E. O. 1967. A Primer on Food, Agriculture, and Public Policy. Random House, New York. 177 pp.

Heimpel, A. M., and T. A. Angus. 1963. Diseases caused by certain sporeforming bacteria. pp. 21-73, Vol. 2, in Insect Pathology, An Advanced Treatise. E. A. Steinhaus, ed. Academic Press, New York.

Helms, D. 1979. Revision and Reversion: Changing Cultural Control Practices for the Cotton Boll Weevil. Paper presented at the Agricultural History Symposium on Science and Technology in Agriculture, Manhattan, Kansas, March 19.

Hightower, J. 1973. Hard Tomatoes, Hard Times. Shenkman, Cambridge. 268 pp.

Hoffmann, C. H. 1970. Alternatives to conventional insecticides for the control of insect pests. Agr. Chem. 25(9): 14-19a, 25(10):19, 21-23, 35.

Holloway, J. K., and C. B. Huffaker. 1952. Insects to control a weed. pp. 135-140 in Insects, The Yearbook of Agriculture 1952. Government Printing Office, Washington.

Howard, L. O. 1930. A History of Applied Entomology (some-
      what anecdotal). Smithsonian Institution, Washington.
      564 pp.

Huffaker, C. B. 1971. Unpublished memo to Proposed Ad-Hoc
      Committee to Consider "The Role of Economic and Systems
      Analysis in the IBP-Biological Control-Crop Ecosystems
      Project. June 25.

Huffaker, C. B., and B. Croft. 1978. Integrated Pest
      Management. Calif. Agric., February. pp. 6-7.

Jones, D. P. 1973. Agricultural entomology. pp. 302-332 in
      History of Entomology. R. F. Smith, T. E. Mittler and
      C. N. Smith, eds. Annual Reviews, Inc., Palo Alto.

Karlson, P. and A. Butenandt. 1959. Pheromones (ecto-
      hormones) in insects. Ann. Rev. Entomol. 4: 39-58.

Kendeigh, S. C. 1947. Bird Population Studies in the
      Coniferous Forest Biome during a Spruce Budworm Out-
      break. Department of Lands and Forests, Ontario,
      Canada. Biological Bulletin #1.

Kennedy, J. S. 1968. The motivation of integrated control.
      J. Appl. Ecol. 4: 492-499.

Knipling, E. F. 1953. The greater hazard: Insects or
      insecticides. J. Econ. Entomol. 46: 1-7.

Knipling, E. F. 1959. Screwworm Eradication: Concepts and
      Research Leading to the Sterile-Male Method.
      Smithsonian Report for 1958: 409-418.

Knipling, E. F. 1962. Introduction. pp. 1-5 in Proceedings
      of Boll Weevil Research Symposium. USDA.

Knipling, E. F. 1966a. The entomologist's arsenal. Bull.
      Entomol. Soc. Amer. 12: 45-51.

Knipling, E. F. 1966b. Some basic principles in insect
      population suppression. Bull. Entomol. Soc. Amer. 12:
      7-15.

Knipling, E. F. 1978. Advances in technology for insect
      population eradication and suppression. Bull. Entomol.
      Soc. Amer. 24: 44-52.

Kuhn, T. S. 1970. The Structure of Scientific Revolutions, 2nd edition. University of Chicago Press, Chicago, 210 pp.

Kuhn, T. S. 1977. The Essential Tension. University of Chicago Press, Chicago. 366 pp.

Lakatos, I., and A. Musgrave, eds. 1970. Criticism and the Growth of Knowledge. The University Press, Cambridge. 282 pp.

Linduska, J. J. 1978. Evaluation of soil systemics for control of Colorado Potato Beetle on tomatoes in Maryland. J. Econ. Entomol. 71: 647-649.

Lyle, C. 1947. Achievements and possibilities in pest eradication. J. Econ. Entomol. 40: 1-8.

Macleod, R. 1977. Changing perspectives in the social history of science. pp. 149-195 in Science, Technology, and Society. Ina Spiegel-Rösing and Derek de Solla Price, eds. Sage Publications, London.

Mason, S. F. 1962. A History of the Sciences. Collier Books, New York. 638 pp.

Meyer, A. S. 1972. Cecropia juvenile hormone, harbinger of a new age in pest control. pp. 317-335 in Insect Juvenile Hormones. Julius J. Menn and Morton Beroza, eds. Academic Press, New York.

Michelbacher, A. E., and O. G. Bacon. 1952. Walnut insect and spider-mite control in northern California. J. Econ. Entomol. 45: 1020-1027.

Murphy, E. F. 1971. Has Nature Any Right to Life. Hastings Law Journal 22: 467-484.

NAS. 1966. Scientific Aspects of Pest Control. NAS, Washington. 470 pp.

NAS. 1975a. Pest Control: An Assessment of Present and Alternative Technologies. NAS, Washington. Vols. 1-5.

NAS. 1975b. Cotton Pest Control. Vol. III in Pest Control: An Assessment of Present and Alternative Technologies. NAS, Washington. 139 pp.

Newsom, L. D. 1974. Pest management: History, current
    status and future progress. pp. 1-18 in Proceedings
    of the Summer Institute on Biological Control of Plant
    Insects and Diseases. F. G. Maxwell and F. A. Harris,
    eds. University Press of Mississippi, Jackson.

Newsom, L. D. 1978. Eradication of plant pests--con.
    Bull. Entomol. Soc. Amer. 24: 35-40.

Nixon, R. M. 1972. Environmental Protection. U. S. House
    of Representatives Document 92-247, Feb. 8. 16 pp.

Noble, D. F. 1977. America by Design. Alfred A. Knopf,
    New York. 384 pp.

Painter, R. H. 1951. Insect Resistance in Crop Plants.
    MacMillan, New York. 520 pp.

Perelman, M. 1977. Farming for Profit in a Hungry World.
    Allanheld, Osmun & Co., Montclair, N. J. 238 pp.

Perkins, J. H. 1978a. Edward Fred Knipling's sterile-male
    technique for control of the screwworm fly. Environ.
    Review 5/78: 19-37.

Perkins, J. H. 1978b. Reshaping technology in wartime:
    The effect of military goals on entomological research
    and insect-control practices. Technology and Culture
    19: 169-186.

Perkins, J. H. In preparation. Boll Weevil Eradication:
    Changing Technologies for Plant Protection and their
    Implications for Public Policy.

Pickett, A. D. 1949. A critique of insect chemical
    control methods. Can. Entomol. 81: 67-76.

PSAC. 1963. Use of Pesticides. Government Printing Office,
    Washington. 25 pp.

Rabb, R. L. 1978. Eradication of plant pests-con. Bull.
    Entomol. Soc. Amer. 24: 40-44.

Ripper, W. E. 1956. Effect of pesticides on balance of
    arthropod populations. Ann. Rev. Entomol. 1: 403-438.

Rudd, R. L. 1964. Pesticides and the Living Landscape.
    The University of Wisconsin Press, Madison. 320 pp.

Scruggs, C. G. 1975. The Peaceful Atom and the Deadly Fly.
    Jenkins Publ. Co., The Pemberton Press, Austin. 311 pp.

Shepard, H. H. 1951. The Chemistry and Action of Insecti-
    cides. McGraw-Hill Book Co., New York. 504 pp.

Smith, E. H. 1964. Pesticides and people. New York State
    Hort. Soc. Proc. 109: 188-193.

Smith, R. F. 1975. The origin of integrated control in
    California--An account of the contributions of C. W.
    Woodworth. Pan-Pac. Entomol. 50: 426-429.

Smith, R. F., J. L. Apple, and D. G. Bottrell. 1976. The
    origins of integrated pest management concepts for
    agricultural crops. pp. 1-16 in Integrated Pest
    Management. J. L. Apple and R. F. Smith, eds. Plenum
    Press, New York.

Smith, R. F., and H. T. Reynolds. 1966. Principles,
    definitions, and scope of integrated pest control.
    pp. 11-17 in Proceedings of the FAO Symposium on
    Integrated Control. FAO, Rome.

Smith, R. F., and R. van den Bosch. 1967. Integrated
    control. pp. 295-340 in Pest Control. W. W. Kilgore
    and R. L. Doutt, eds. Academic Press, New York.

Special Study Committee on Boll Weevil Eradication. 1969.
    Selection of locations for pilot boll weevil eradica-
    tion experiments [National Cotton Council, Memphis (?)].

Staub, R. W., and A. C. Davis. 1978. Onion maggot:
    Evaluation of insecticides for production of onions in
    muck soils. J. Econ. Entomol. 71: 684-686.

Steele, J. E. 1976. Hormonal control of metabolism in
    insects. pp. 239-323 in Advances in Insect Physiology.
    J. E. Treherne, M. J. Berridge and V. B. Wigglesworth,
    eds. Vol. 12.

Steinhaus, E. A. 1951. Possible use of Bacillus thuringi-
    ensis Berliner as an aid in the biological control of
    the alfalfa caterpillar. Hilgardia 20: 359-381.

Stern, V. M., R. F. Smith, R. van den Bosch, and K. S. Hagen.
    1959. The integrated control concept. Hilgardia 29:
    81-101.

U. S. Congress. 1950. House Resolution 323 (81:1), passed
    June 20.

U. S. Congress. 1952. House Select Committee to Investigate
    the Use of Chemicals in Foods and Cosmetics, Food,
    H. Rpt. 82: 2356, June 30.

U. S. Congress. 1965. Agricultural Appropriations for
    1966, Hearings, Part 1, Senate Committee on Appropria-
    tions. Government Printing Office, Washington.

U. S. Congress. 1966a. Agricultural Appropriations for
    Fiscal Year 1967, Hearings, Part 1, Senate Committee
    on Appropriations. Government Printing Office,
    Washington.

U. S. Congress. 1966b. Pesticides and Public Policy.
    Senate Committee on Government Operations, 89:2,
    Report No. 1379. 86 pp.

USDA. 1962. Comment on Rachel Carson's Articles in the
    New Yorker. 2 pp.

U. S. Department of Health, Education and Welfare. 1969.
    Report of the Secretary's Commission on Pesticides and
    their Relationship to Environmental Health. Government
    Printing Office, Washington. 677 pp.

US EPA. 1977. Suspended and Cancelled Pesticides. EPA,
    Washington. 16 pp.

van den Bosch, R. 1978. The Pesticide Conspiracy.
    Doubleday and Co., Inc., Garden City. 226 pp.

Whorton, J. 1974. Before Silent Spring. Princeton
    University Press, Princeton. 288 pp.

Wigglesworth, V. B. 1970. Insect Hormones. W. H. Freeman
    and Co., San Francisco. 159 pp.

Wilson, F., and C. B. Huffaker. 1976. The philosophy,
    scope, and importance of biological control. pp. 3-15
    in Theory and Practice of Biological Control. C. B.
    Huffaker and P. S. Messenger, eds. Academic Press,
    New York.

3. The Economic Milieu of Pest Control: Have Past Priorities Changed?

Introduction

Secretary of Agriculture, Charles Brannan, in a foreword to the Yearbook of Agriculture in 1952 made the following statement:

"We dare not think of any knowledge-least of all knowledge of living things-as static, fixed or finished."

He made this statement in the foreword of a book on insects after noting that even though the science of entomology had made great progress in the two decades prior to 1952, the problems caused by insects seemed to be even bigger than ever.

Some 26 years later, these observations still carry meaning for people knowledgeable about the problems of pest control in agriculture. The combination of a constant pressure by pests on agricultural production, concerns for producing enough food and the concern for the effects of chemical pest control agents on humans and the environment constitute ample evidence to

---

*Research on which this paper is based was supported by the Missouri Agricultural Experiment Station and by a grant from Resources for the Future, Inc. This paper has benefited from comments by David Pimentel and John Perkins on an earlier draft.

Table 1. Selected Characteristics of U.S. Agriculture
          1952 and 1975.

Characteristics	Units	1952	1975
Farm population	10^6 persons	24.2	8.8
Farm population as percent of total population	%	15.5	4.2
Number of farms	10^6 farms	5.4	2.8[2]
Average farm size	acres	216.0	440.0
Cropland harvested	10^6 acres	387.0[1]	303.0[2]
Irrigated land	10^6 acres	29.5	41.2[2]
Cotton acreage	10^6 acres	25.9	12.3[3]
Corn acreage	10^6 acres	81.0	77.9
Soybean acreage	10^6 acres	14.3	53.6
Cotton acreage treated for insects	10^6 acres	13.0	7.5[3]
Small grain acreage sprayed for weeds	10^6 acres	17.0	33.7[3]
Nitrogen use	10^3 tons	1637.0	8607.7
Foreclosure and bankruptcy	No./1000 farms	1.5	1.5

Source: U.S. Department of Agriculture, 1954; 1976a
          U.S. Department of Commerce, 1978
          Andrilenas, 1975

[1] Data for 1950

[2] Data for 1974

[3] Data for 1971

conclude that pest control problems are far from being solved once and for all. During those 26 years, however, many things have changed, which suggest that new approaches to pest control problems may be needed. The technology applied to agriculture has changed, the pattern of farm ownership and operation has changed, consumer tastes have changed, and the domestic and international economic climates have changed dramatically since the close of the Korean War.

It is the purpose of this paper to provide a detailed description of the changes in the economic milieu that surrounds agriculture and pest control technology. This description will provide a basis for (1) explaining the development of agricultural pest control and (2) providing insight into actions that are needed to improve pest control to meet social objectives.

Changes in American Agriculture

The observation of Secretary Brannan mentioned above, indicated that agriculture was a fast changing activity with significant pest control problems that seemed to be growing. A brief perusal of some data on U.S. agriculture indicates that these observations were accurate. Due to various forces, both inside and outside agriculture, the nature and scope of farming has changed significantly since 1952.

Table 1 displays selected characteristics of U.S. agriculture with comparisons between 1952 and about 1975. First, the number of people living on farms has declined by about 60% to a point where only one person in 25 now is classed as farm population. Farm numbers have been cut in half since 1952 and farm size has doubled.

While the cropland harvested has declined by about 20%, irrigated land has increased by about 40% representing a significant increase in the input intensity on the land that is irrigated. Acreages of three important crops, corn, cotton, and soybeans provide an interesting comparison over the 1952-75 period. Cotton acreage declined by about 53% due to eocnomic pressures from demand. As a result, many areas in the so-called "cotton south" reduced acreages by extremely large amounts. States such as Virginia and North Carolina grow almost no cotton now. Corn acreage has changed very little since 1952, while soybean acreage has increased by a

factor of 3.7. The increase in soybean production has
been in response to a strong domestic and international
demand for oil and protein. Soybeans displaced cotton
in the south and probably hay in the north central U.S.
as livestock technology came to rely less on pasture and
roughage and more on concentrates. In addition, the
development of herbicides made weed control much more
effective resulting in increased yields.

Farmers' use of chemical technology has changed as
well. In 1952 about half of the cotton acreage was
chemically treated for insects. In 1971, 61% of the
cotton acreage was treated with chemical insecticides
of one kind or another. In 1952, 12% of small grain
acreage was sprayed with herbicides for weed control, while
in 1971, 37% of the small grain acreage was so treated.
Finally, the use of nitrogen fertilizer, which has been
important in increasing yields, has increased by more
than a factor of 5.

These data demonstrate that, indeed the agriculture
of the present is very different from the industry that
existed during the Korean War. Farmers spent over 1
billion dollars for pesticides in 1971, 80% more than they
spent in 1966. Presently more than 1 billion pounds of
active chemical ingredients are used by farmers, about
twice the amount used in 1964. Chemical technology has
been rapidly applied to agriculture during the last 20
years, especially in the control of insects, weeds and
plant and animal diseases.

The Economic Context

Producing agriculture is an industry made up of 2.8
million individual firms. Most of the labor used by those
firms is supplied by the owners of the firms or by
persons within their family. It is an industry of price
takers in the market. That is, it is not organized so
that the producers can administer the prices which they
receive for their products, neither can the firms in any
way control the prices paid for inputs. It is an industry
of what economists call atomistic competition.

In this setting of atomistic competititon, one finds
also a series of national policies that have had effects
on the business of farming. First, since the close of
the Great Depression it has been a national goal to
facilitate economic growth and development by applying
technology to agriculture in an effort to reduce the

resources devoted to agriculture, especially labor. The consequence has been the reduction in farm population and the number of farms shown in Table 1. Capital has been substituted for labor in a dramatic way in farming. Second, it has been a policy and a matter of national pride to reduce the amount of consumers' incomes spent for food, thereby releasing more income for investment and the purchase of industrial goods. This policy has also been successful. In 1951, consumers spent between 25 and 30% of their disposable incomes for food. In 1960, the estimate was 21% and in 1976 the comparable figure was 19% (U.S. Department of Agriculture, 1976b; U.S. Department of Commerce, 1953; 1976). The important factor in the achievement of the "cheap food" policy has been the continued expansion of the supply of food.

As a result of these two policies, farming has become dependent on the industrial sector for large amounts of the inputs, placing demands on the farm business to generate cash income to pay for the purchased inputs. In 1952, production expenses were 63% of gross farm income. By comparison, production expenses were 75% of gross farm income in 1975 (U.S. Department of Agriculture, 1954; 1976a).

The pressure on farmers to compete in this sort of environment has strengthened the demand for technology which while increasing the relative cash costs of production, holds the expectation that unit costs of production will be reduced relative to the prices received for products. There is considerable evidence that the demand for chemical pest control has increased because it was productive. Studies by the author were done based on data from the mid-1960's (Headley, 1968, 1970). These studies estimated the aggregate production function for U.S. agriculture and estimated the contribution of pesticides to agricultural productivity. The results of one of these studies (1968) showed that pesticides as a group had a marginal value product of $4 per $1 of incremental expenditure. A later study (1972) showed varying productivity for the different kinds of chemical pesticides by regions of the country. In most regions, the marginal value products of both herbicides and insecticides were estimated at several dollars per pound of active ingredients used. Regions with higher use intensity demonstrated lower marginal productivities, a finding consistent with economic theory.

The empirical work on pesticide productivity supports the general conclusion that pests are a source of economic damage and that farmers have found chemical pesticides to be profitable in attempting to control that damage.   It is argued that this political and economic climate provides, in large part, an explanation for the dramatic increase in the use of commercial fertilizer and chemical pest control products by farmers.

The Decision-Making Framework
for Pest Control

As data from the U.S. Department of Agriculture indicate, about half or over one million farmers are applying chemical pesticides as a means of pest control on crops (Andrilenas, 1975).   This means that, at a minimum, over one million decisions are made annually about the application of chemicals.   As chemical pest control has developed, this technology is adapted to numerous independent decisions.   Farmers perceive their pest problems and can act on them quickly and unilaterally. It is a technology that fits the atomistic nature of the farming industry.   The producers of chemical materials can identify their markets and advertise to promote the sales of their products with the target of that promotion being the individual farmer-the customer.

This sort of decision making framework does not lend itself necessarily to either (a) the consideration of technical spillover effects on parties who are not directly involved in pest control decisions or to (b) the consideration of the broader long-term agro-ecosystem effects.   It is a decision framework based on short-term private economic benefits and costs.   Consideration of the external benefits and costs of particular pest control decisions cannot be given nor is it likely that farmers will consider the effects of a particular decision on pest resistance in the future, or the development of secondary pests.

There is the case of cotton growers in the lower Rio Grande Valley who, due to secondary pests, found it necessary to move away from a program of control through a rigid schedule to chemical insecticide treatments.   They have modified their programs to a "treat as needed" program combined with cultural practices and improved varieties.   However, this was not done in anticipation of secondary pest problems.   These growers had no choice if they wished to continue in cotton production (National Academy of Sciences, 1975).

It is a natural result of the institutional setting that the public has become involved in pest control decisions through the Federal Environmental Pest Control Act of 1972 (FEPCA) and its precursor, the Federal Insecticide, Fungicide and Rodenticide Act of 1947 (FIFRA). In fact, pesticide laws date back in history to 1898 with the passage of a law in New York State to regulate paris green (Lemmon, 1952). These laws represent attempts by society to protect not only the users of pesticides, but also consumers and the natural environment from adverse effects. They represent a recognition of the fact that the market for pesticidal chemicals fails to incorporate the social values that are involved in pesticide decisions. In the jargon of economics, this phenomenon is referred to as imperfections in the market. These imperfections are due to a lack of knowledge-users don't really understand the products they use-and to the fact that there is no way to provide incentives for the user to consider those benefits and costs, which are not measured by the cost of the chemicals applied or by the prices of the farm products produced (Headley and Lewis, 1967).

Pest Control Problems

Previous sections of this paper have alluded to problems with pest control in modern agriculture. These problems are: (1) the need to develop new methods to deal with pest resistance developed to previous chemical materials, (2) the need to develop controls for secondary pests made necessary by the reductions in beneficial species as the result of previous use of wide spectrum pesticides, (3) the need to reduce hazards to farm workers and other humans due to the use of chemical compounds that are either highly toxic or problems of a more chronic toxicity or both and (4) the need to reduce hazards to non-target species in the natural environment such as fish and wildlife.

Problems of Pest Resistance

As soon as chemicals began to be widely used in U.S. agriculture to control pests, especially arthropods, pest populations came under heavy chemical pressure and resistance began to develop. It was first noticed in the U.S. in San Jose scale control about 1914 (Porter, 1952). DDT resistance in flies was reported in Italy and Sweden in 1947 (Bruce, 1952). Mosquito resistance showed up at about the same time. Many other cases of resistance have been documented not only among arthropods, but also among the fungi (Reynolds et al., 1975; Schuntner et al., 1968; Georgopolous, 1977; Pate and Vinsora, 1968).

The economic importance of chemical resistance is that: (a) it increases pest losses because the degree of control declines over time, (b) it increases control costs due to the requirement of extra treatments and (c) it increases control costs because of new investment required to develop replacement compounds (Headley, 1972).

In a system where individual farmers make pest control decisions on a short-run basis there is always the danger that chemicals will be applied enhancing resistance when the long-run benefits do not justify it either to the individual farmer or to society as a whole. What happens in an economic sense is that the life of the capital asset, the susceptible gene pool is shortened due to an accelerated depreciation (Hueth and Regev, 1974). The only way to correct this difficulty is to reduce the myopia in pest control strategies. To do this requires both a higher level of knowledge regarding economic thresholds and a method of extending the length of the planning horizon in making pest control decisions. In order to bring this higher level of management to pest control, one must depart from the past decision system of individual actions and methods adapted to that setting and move more toward a setting where decisions are made in a broader geographical and time framework requiring more coordinated action by growers. The political and economic dimensions of this broader decision framework need to be examined since little is known about how to proceed.

Secondary Pest Problems

There is evidence to indicate that some of the pest problems which now exist in agriculture have been induced by the use of broad spectrum chemicals used on certain key pests such as boll weevil and lygus on cotton in the south and California respectively (Reynolds et al., 1975).

The numerous cases of secondary pests which have appeared are evidence that the methods used for dealing with pests have tended to be counterproductive in the long run. Resources are now required to control pests that were previously controlled by natural means. Farmers and society are therefore now bearing the costs of previous pest control strategies. Such a situation is surely the result of the decision framework where unilateral control decisions are made with non-selective methods and without appropriate information or incentives to remove the myopia from those decisions.

Problems of Hazards to Humans

Since chemicals used in pest control exhibit varying degrees of toxicity to warm-blooded species, there is always danger of a hazard to farm workers, to people in the community where chemicals are used, and to consumers of the products on which the chemicals are applied. Early in the history of chemical pest control, the principal concern was for acute toxicity to humans and farm animals. The introduction of DDT was hailed as a remarkable breakthrough because of its extremely low mammalian toxicity. That idea has since been found to be oversimplified.

The discovery of chlorinated hydrocarbons in animal fat and their dispersal throughout the food chain plus the linking of DDT to cancer in laboratory animals led to the eventual banning of DDT and other chlorinated hydrocarbon compounds. Further, the discovery of the oncogenic properties of pesticidal chemicals as well as mutagenic and teratogenic properties led eventually to the full-scale review of all registered chemical pesticides by the U.S. Environmental Protection Agency. Again one sees the result of a decision framework based on individual decisions that have the appearance of optimality when in fact there may be significant technical external effects that are not known and therefore cannot be considered as a part of the decision. Governmental regulation has been relied on to deal with this market imperfection up to this point.

Problems of Hazards to the Environment

Effects of chemicals on the natural environment have also been identified. Fish, birds, and other nontarget species are affected by chemicals applied to agricultural crops. How important these external effects are is in many cases difficult to determine. Some of these species, such as birds, are perceived as valuable because of their roles in natural control of pests. Others are seen as important to other species as a source of food. Still others, such as fish, birds, and some mammals are valued purely for their aesthetic value to many members of society. These are effects that the market for chemical pesticides finds difficult, if not impossible, to measure and include in the myriad of pest control decisions made by farmers, homeowners, and others. Again, regulation has been adopted as the method of controlling these external effects.

There are many shortcomings of the regulation approach. It
has been successful in dealing with the grossest of environmental hazards, but is not well adapted to deal with the more
subtle biological, chemical, and socio-economic interactions.

Alternative Control Strategies

It is apparent that the pest control strategies that
have been followed need to be improved. Evidence available
indicates that pest problems are as severe, if not more so,
than there were 20 years ago. It is also apparent that at
least for the foreseeable future, the need for food and fiber
by the world will be such that capital and land-intensive
agriculture will be necessary to prevent hunger. There is
a need to change the strategies to minimize the problems
associated with chemicals, chosen and applied by individuals,
as one of the principal defenses against insect, weeds,
pathogens, and other pests. The development of such strategies will be difficult because of the wide range of social,
economic, and political values involved.

Alternatives Available

To deal with the shortcomings of the chemical pest control strategy, two major ideas are being examined. One
approach is to substitute, in part, biological and cultural
controls for chemicals and the second is to abandon the philosophical concept of reducing pest populations to an absolute
minimum in either a short- or long-term planning horizon and
to embrace the concept of the management of pest populations
at some level determined to be economically justified.

It has been shown that through the use of parasites,
predators, attractants, growth regulators, pathogens, and
host plant resistance, that at least certain pest species
populations can be controlled. It is also well known that
cultural methods such as crop rotations, timing of planting,
cultivation, and the management of plant residues and alter-
nate hosts can control certain pests.

Biological and cultural controls, even though used extensively, are faced with two major problems when placed
in the commercial agriculture context. First, the
research information needs are great. To apply new additional biological and cultural methods requires an understanding of the basic biology of the pest including its
life cycle and its natural enemies. This is a time-
consuming and expensive capital investment process. Second,
there are marketing and distribution problems. Were

biological methods commercially saleable, the capital
investment needed for their development and distribution
would be forthcoming from private enterprise concerns
motivated by the profit incentive. However, such is not
usually the case. While it is possible to rear and sell
natural enemies, and in fact it is being done, much of
what is involved in biological and cultural methods is the
development of information. The economics of commercial
information is complicated by problems of maintaining
proprietary control so that the developer can recover the
necessary economic return necessary for its generation.

For the reasons just cited, the outlook for the rapid
replacement of a pest control strategy based principally
on chemicals is not bright at least for the next 10-15
years. A research study by the author, just concluded
(Headley, 1978) surveyed a panel of U.S. agricultural
experts in extension and research to obtain estimates of
the probable importance of the spectrum of pest control
techniques over the next 15 years. These responses are
presented in Table 2. Their conclusions were that chemicals
will continue to be of major importance, but that the use
of insecticides will decline and herbicide use will
increase. Among the biological methods, they believe
resistant varieties to be the most promising for grains
and soybeans with cultural methods being relatively minor.
The experts believe that the use of bacteria will increase,
although it will remain a minor technique. The use of predators and parasites will change little, and the outlook for
the more exotic measures such as viruses, pheromones, and pest
genetics is not hopeful. Certainly the work on citrus,
cotton and selected vegetable and orchard crops makes the
outlook there more promising, but there is still a long way
to go to get the methods widely applied.

A technology assessment by a large consulting firm
supports this general assessment (Lawless and von Rümker,
1976). This study cites as disadvantages, the unsuit-
ablility of certain biological agents to use by individuals and special training or education required which is
not possessed by the average producer. They point out
that the selectivity of many methods, while environmentally
desirable, will handicap them due to the restricted
size of the market relative to the developmental costs.

Cultural methods can and are being used by farmers.
These amount to management techniques, which most farmers
are capable of applying if they are demonstrated effective
and if benefit-cost ratios show them to be profitable.

Table 2. Estimated Importance of Pest Control Methods for
Grain Crops and Soybeans, U.S. Agriculture
1978-1992.

Pest Control Technique	Probable Use Over Next 15 Years	Trend in Use[1]
Chemical Poisons		
Insecticides	major	-
Herbicides	major	+
Mechanical Methods	minor	-
Biological Methods		
Parasites and Predators[2]	minor	0
Bacteria	minor	+
Viruses	not significant	+
Pheromones	not significant	0
Resistant Varieties	major	+
Pest Genetics	minor	-
Cultural Methods		
Crop Rotations	minor	-
Trap Crops	minor	0

Source: A summary of responses from 39 U.S. agricultural
extension and research workers, 1977.

[1] "+", "-", or "0" means a trend that is increasing,
decreasing or unchanged respectively.

[2] Parasites and predators refers to their application of
parasites and predators. Naturally occurring parasites
and predators have been and will continue to be
important in insect control.

They do add to costs of production either through the use
of more labor and capital or through production lost as in
the case of rotations involving crops with lower profit
margins. This points directly to the way agriculture has
adjusted to the recent economic pressures brought by the
policy of industrialization leading to higher labor costs,
the force of international demand for crops such as soybeans
and the policy of cheap food.

As a compromise between complete dependence on
chemicals or on application of natural enemies, the
concept of integrated pest management has been developed
to combine the use of better technical management skills
by farmers with a mix of chemical, biological and cultural
methods. In this way it is hoped that the reliance on
chemicals can be reduced and the problems associated with
chemicals, therefore reduced (National Academy of Sciences,
1969).

Research on integrated pest management is going
forward. The Federal Extension Service of the U.S. Department of Agriculture has been sponsoring educational
programs for farmers and their advisors to improve their
skills. Both the public and the private sectors of the
economy have roles to play in the development of integrated
control.

Much of the research needed as a foundation is basic
biological research. Pest management means the maintenance of pest populations at levels that are greater than
zero, but below levels where unacceptable economic damage
to production exists. This requires more knowledge than we
currently have about the pest-enemy-host relationships.
Most of this research will become the responsibility of the
public sector.

Chemicals are most effective in quickly reducing
pest populations where explosions occur. Individual
farmers will continue to need tools such as these to
deal with the inability to fine tune an agro-ecosystem
everywhere all of the time. The private sector currently
has the responsibility for the development of chemicals
that are compatible with the various biological and
cultural methods that are a part of the integrated
approach. Whether this responsibility can be fulfilled
remains to be seen.

Implications for the System

The purpose of this paper was to describe the changes
in agriculture and agricultural pest control in response
to its economic milieu and to provide insights needed to
improve pest control to meet social objectives. These
social objectives continue to be: (a) an economically
viable agriculture consisting of a large number of small
units, (b) a minimization of the labor and other resources
devoted to food and fiber production, (c) ample nutritious
food which can be purchased with the incomes of consumers
and (d) an environment that is a safe and aesthetically
pleasing place in which to live. These are challenging
objectives.

In order to meet them, changes in the institutional
framework will be needed. In order to deal with the very
real pest problems that exist, new ways will need to be
found to maintain the productivity of agriculture.

It has been argued here that due to the imperfections
of markets, atomistic competition in production agriculture
will not produce a socially acceptable level or kind of
pest control. The implication is that some method must be
found to deal with these market imperfections. The
improvements must adjust for the external effects of the
use of chemical pesticides and adjust for short-term
planning horizons inherent in the decision framework
where pest control decisions are made.

The specific implication is the enlargement of the
role that the public at large plays in the development
and application of pest control for agriculture. It has
been found that governmental regulation of pesticides is
inadequate to the task. While useful for the most gross
kinds of problems, such as acute and chronic toxicity and
short-term efficacy, it cannot, in an economically
efficient manner, deal with the other problems that plague
pest control. Something in addition to the current
regulations is therefore needed.

The need for research and education to cope with the
problems of minimizing the resources devoted to pest
control, to find ways to make better use of the natural
environment, to maintain ecological stability and to
determine the essentiality of various technologies calls
for a larger public investment in research and development in pest control. Federal and state agencies and the
university community need to provide greater leadership

to see that this job is done. Agricultural colleges, which are closest to and better able to understand the technical and economic situation of the farmers of their region, need to be given the specific responsibility to do the research needed to develop and apply integrated pest management for their area. In the development of selective chemicals necessary for use with biological methods, policies are needed to produce and market chemicals developed with public funds to further the objectives of integrated control. Whether this should be done by franchising private concerns to produce and market chemicals or whether a T.V.A.-like approach to pesticides might be better, needs to be studied.

At the farm level there is a need to design institutions to first broaden the area to which pest control decisions apply and second, as a necessary complement, to provide ways for the pooling of the risks taken by individual farmers under this broad area approach to pest control. The establishment of pest control districts, perhaps conforming to counties, coupled with a scheme of crop insurance could provide a setting where pest control decisions are made that consider both external effects and longer term concerns for ecological stability. At the same time, individual farmers could be protected from any adverse effect that such decisions might have for them.

There is much yet to be done. There are many questions for which there are no answers. Courage and imagination on the part of scientists, politicians and the business community are very much needed if the old priorities and policies are not to become those of the future.

References

Andrilenas, P. 1975. Farmers' Use of Pesticides in 1971. Agr. Econ. Rep. No. 268. U.S. Department of Agriculture, Washington, D.C.

Bruce, W.N. 1952. Insecticides and flies. In Insects-The Yearbook of Agriculture. U.S. Congress House Document No. 413, Washington, D.C. 780 pp.

Georgopolous, S.G. 1977. Pathogens become resistant to chemicals. In Plant Disease: An Advanced Treatise. J.G. Horsfall and E.B. Cowling, eds. Academic Press, New York. 465 pp.

Headley, J.C. and J.N. Lewis. 1967. The Pesticide Problem: An Economic Approach to Public Policy. Johns Hopkins, Baltimore. 141 pp.

Headley, J.C. 1968. Estimating the productivity of agricultural pesticides. Am. J. Agr. Econ. 50(1):13-23.

Headley, J.C. 1970. Productivity of Agricultural Pesticides. Proceedings of a U.S. Department of Agriculture Symposium: Research on Pesticides for Policy Decisionmaking.

Headley, J.C. 1972. Economics of agricultural pest control. In Annual Review of Entomology. R.F. Smith, T.E. Mittler, and C.N. Smith, eds. 17:555 pp.

Headley, J.C. 1978. Pest control as a production constraint for grain crops and soybeans in the U.S. to 1990. Unpublished manuscript, University of Missouri (Columbia).

Hueth, D. and U. Regev. 1974. Optimal agricultural pest management with increasing pest resistance. Am. J. Agr. Econ. 56:543-552.

Lawless, E.W. and R. von Rümker. 1976. A Technology Assessment of Biological Substitutes for Chemical Pesticides. Midwest Research Institute, Kansas City, Missouri. Draft report, 503 pp.

Lemmon, A.B. 1952. State pesticide laws. In Insects-The Yearbook of Agriculture. U.S. Congress House Document No. 413. Washington, D.C. 780 pp.

National Academy of Sciences. 1969. Insect Pest Manage-
     ment and Control. Publication 1695. Washington, D.C.
     508 pp.

National Academy of Sciences. 1975. Pest Control: An
     Assessment of Present and Alternative Technologies.
     Vol. III. Cotton Pest Control. 139 pp.

Pate, T.L. and S.B. Vinsora. 1968. Evidence of non-specific
     type resistance to insecticides by a resistant strain
     of the tobacco budworm. J. Econ. Entomol. 61(11):35-37.

Porter, B.A. 1952. Insects are harder to kill. In Insects-
     The Yearbook of Agriculture. U.S. Congress House Docu-
     ment No. 413. Washington, D.C. 780 pp.

Reynolds, H.T., P.L. Adkisson and R.F. Smith. 1975. Cotton
     insect pest management. In Introduction to Pest Manage-
     ment. R. Metcalf and W. Luckmann, eds. Wiley, New
     York. 587 pp.

Schuntner, C.A., W.J. Roulston and H.J. Schnitzerling. 1968.
     A mechanism of resistance to organophosphorous acaricides
     in a strain of the cattle tick, Boophilus microplus.
     Austral. J. Biol. Sci. 21:91-109.

U.S. Department of Agriculture. 1954. Agricultural
     Statistics. U.S. Government Printing Office, Washington,
     D.C. 607 pp.

U.S. Department of Agriculture. 1976a. Agricultural
     Statistics. U.S. Government Printing Office, Washington,
     D.C. 607 pp.

U.S. Department of Agriculture. 1976b. National Food
     Situation-February. Econ. Res. Serv., Washington,
     D.C. 51 pp.

U.S. Department of Commerce. 1953. Survey of Current
     Business. Vol. 33, No. 1. Bureau of Economic Analysis,
     Washington, D.C. 16 pp.

U.S. Department of Commerce. 1976. Survey of Current Busi-
     ness. Vol. 58, No. 2. Bureau of Economic Analysis,
     Washington, D.C. 16 pp.

U.S. Department of Commerce. 1978. Farms: Number, Acreage,
     Value of Land and Buildings, Land Use, Size of Farm,
     Farm Debt. 1974 Census of Agriculture, Vol. 2, pt. 2.
     Bureau of the Census, Washington, D.C.

_David Pimentel, David Andow,
David Gallahan, Ilse Schreiner, Todd E. Thompson,
Rada Dyson-Hudson, Stuart Neil Jacobson,
Mary Ann Irish, Susan F. Kroop,
Anne M. Moss, Michael D. Shepard, Billy G. Vinzant_

4. Pesticides: Environmental and Social Costs

Abstract

The indirect costs of pesticide use in the United
States are estimated to be nearly $1 billion dollars.
These include large costs from human exposure to pesticides,
increased pest control costs on crops, crop pollination
problems, and pollinator losses (about 70% of the costs),
and also costs from livestock, crop, fish, and wildlife
losses, and government expenditures. Although several
costs are unmeasured, this estimate serves to underscore
the importance of the qualitative unmeasurable costs in
determining pesticide use policies. More effective use of
pesticides is encouraged and the transferral of the indirect
costs of pesticide use to those who reap the benefits is
advocated.

Introduction

Pesticides are an important means of pest control in
the United States (USDA, 1975a; Pimentel et al., 1978a).
Increasing amounts of pesticide are being used (Figure 1)
and in 1978 about 800 million pounds of pesticides were
applied to crop lands. An additional 200 million pounds
were used by homeowners and state and federal agencies for
pest control (Berry, 1979). Some of the latter insecticide
was used to prevent disease spread by vector insects as
well as to eliminate nuisance pests (NAS, 1975). Although
no data are available, probably some human lives were saved
by controlling disease vectors.

The cost of applying pesticides to the 20% of crop
lands treated (USDA, 1975a) is $2.2 billion annually

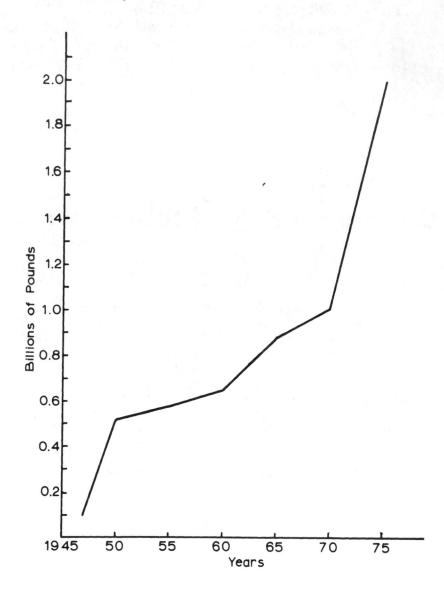

Figure 1. Estimated amount of pesticide produced in the United States (USDA, 1971; Fowler and Mahan, 1975).

(Pimentel et al., 1978a). This price includes materials,
machinery, and labor used for application, and extra treatments required because of pesticide resistance and losses
of natural enemies of pest organisms. This investment
prevents crop losses worth $8.7 billion, or 9% of current
production (Pimentel et al., 1978a). Including the treatment costs for all pests, the total direct cost for pesticides
is $2.8 billion and total benefits $10.9 billion.

This estimate of the cost of pesticides excludes most
of the indirect costs. These indirect or environmental and
social costs must be assessed to facilitate the formulation
of an effective policy of pesticide usage. These environmental and social costs include primary losses attributable
to the use of pesticides. Examples of these would be persons killed or made ill while applying pesticides, herbicide
drift damage to crops and gardens, and honey bee kills.
Secondary losses might include, for example, the cost of
renting colonies of honey bees to overcome pesticide-
induced shortage of pollinators, or loss of income to
commercial fishermen prevented from fishing in water bodies
contaminated by insecticides. Finally, there is a cost
associated with governmental regulation designed to prevent
such damage.

We have attempted to investigate and evaluate the
available data on the indirect costs that result from
pesticide use in this country. Our estimate will necessarily be less than complete. Some of the environmental or
social groups affected have been poorly investigated and
little quantitative data are available. Furthermore, some
losses cannot be evaluated in terms of dollars. Human
lives can be included in this category (a minimum evaluation
was estimated) as can various aesthetic losses such as
reductions of bald eagle and peregrine falcon populations
by DDT and other insecticides (Pimentel, 1971; Stickel,
1973; Edwards, 1973; Brown, 1978a). In addition, we did not
attempt to investigate the distribution of the indirect
costs among different sectors of the population, a factor
which will certainly influence economic decisions and
governmental policies.

One useful distinction in how these costs are borne is
between indirect and external costs. Indirect costs are
defined as any loss due to the effects of pesticides on
nontarget organisms. External costs are less inclusive than
indirect costs, and are only those costs not borne solely
by the individual actually applying the chemicals. For
example, increased crop control costs due to harming beneficial insects and insurance costs for pesticide applicators

are indirect costs borne by the farmer. However, costs such
as damage to other farmers' fields due to drifting herbicides
are external costs.

In this preliminary assessment, we include analyses of
the costs due to: human pesticide poisonings and fatalities;
livestock and livestock product losses; increased control
expenses resulting from pesticide-related destruction of
natural enemies and pesticide resistance; crop pollination
problems and honey bee losses; crop and crop product losses;
fish and wildlife losses; and governmental expenditures to
reduce environmental and social costs resulting from pesticide use.

Costs of Pesticide Exposure to Humans

Undoubtedly, human pesticide poisonings are the highest
price paid for pesticide use. Unfortunately, it is impossible to measure these social costs accurately, because no
one can place an acceptable monetary value on a human life
and extended suffering from chronic illness. We have
attempted to value pesticide-caused illness and death with
standard economic methods, and thereby establish a minimal
estimate for the costs of human pesticide poisonings.

People are exposed to pesticides in a variety of contexts. For example, minute quantities of pesticides are
consumed daily in food and water. About 50% of foods
sampled by the FDA contain detectable levels of pesticides
(Duggan and Duggan, 1973). Food is probably the major
source of low-level chronic exposure to pesticides, but
pesticides may also be absorbed from drinking water, from
contaminated air, and through contact exposure with the
skin (Feldman and Maibach, 1970; Starr and Clifford, 1971;
Stanley et al., 1971; Keil et al., 1972a). As a result of
this chronic exposure, pesticide residues are commonly
found in human tissues. Virtually everyone in the United
States harbors some pesticide residue, averaging 6 ppm in
fatty tissues (Kutz et al., 1977). Even the very young
have detectable residues because human milk and some cow
milk contain pesticide residues (Kutz et al., 1977) and
pesticides may cross the placental barrier (O'Leary et al.,
1970).

A number of subpopulations in the United States are
exposed to higher concentrations of pesticides. These
include: pesticide applicators, flaggers, tank loaders,
farm and field workers, and industrial chemical workers
(Davies et al., 1973; Wolfe, 1976; Milby, 1976; Wicker,
1976). Constantly coming in contact with pesticides, these

groups not only have a higher chronic level of exposure, but they also have an increased chance of acute exposure resulting in immediate illness or death. Accidental acute exposure may also occur in the home as well as on the job. In addition, some of these groups are exposed to special hazards. Pesticides, by altering reaction time and fogging the brain, may increase the chance of an airplane crash for aircraft applicators. In 1976, there were 174 airplane crashes involving pesticide applicators, of which 11 were fatal (NTSB, 1977). Since the cause of the accident is often indeterminate, this type of accident is not counted in our estimate.

In spite of the fact that pesticide residues in humans are ubiquitous, their epidemiological effects have not been well documented (HEW, 1969; Goulding, 1969; NAS, 1975; Barnes, 1976). However, a variety of effects do occur (Figure 2). They may become apparent immediately after exposure or may be delayed. They may only be temporary and subject to treatment, or they may last a long time. Since methods are not available to detect any delayed temporary effects, if they exist they will have to be disregarded.

Pesticides can cause electroencephalogram changes (Metcalf and Holmes, 1969), a variety of neurological alterations (Dille and Smith, 1964; Jenkins and Toole, 1964; Metcalf and Holmes, 1969), psychiatric sequalae (Durham et al., 1965; Stoller et al., 1965; Tabershaw and Cooper, 1966; West, 1968; Metcalf and Holmes, 1969), and may induce parkinsonism (Davis et al., 1978) as well as epilepsy (Nag et al., 1977). Pesticide exposure has been correlated with hypertension (Radomski et al., 1968; Sandifer and Keil, 1971), high blood cholesterol and serum vitamin A concentrations (Sandifer and Keil, 1971; Keil et al., 1972b; Carlson and Kolmodin-Hedman, 1972, 1977), and with cardiovascular disease (Gumennyi and Tkach, 1976). Some pesticides can reduce fertility and may even cause sterility (Whorton et al., 1977; Potashnik et al., 1978; Wheater 1978; Scott, 1978). Other effects are: general blood dyserasias (Best, 1963; Mengle et al., 1966; Takahasi et al., 1978), allergy sensitivity (Milby and Epstein, 1964; Nater and Gooskens, 1976) and possibly liver disease (Cassarett et al., 1968; Radomski et al., 1968; Komarova, 1976; Kim et al., 1977). The magnitude of the delayed effects from acute exposure in occupational groups may just now be appearing (Bidstrup et al., 1953; Fisher, 1977; Davis et al., 1978).

A number of pesticides have been implicated as teratogens in several laboratory organisms, but data on humans

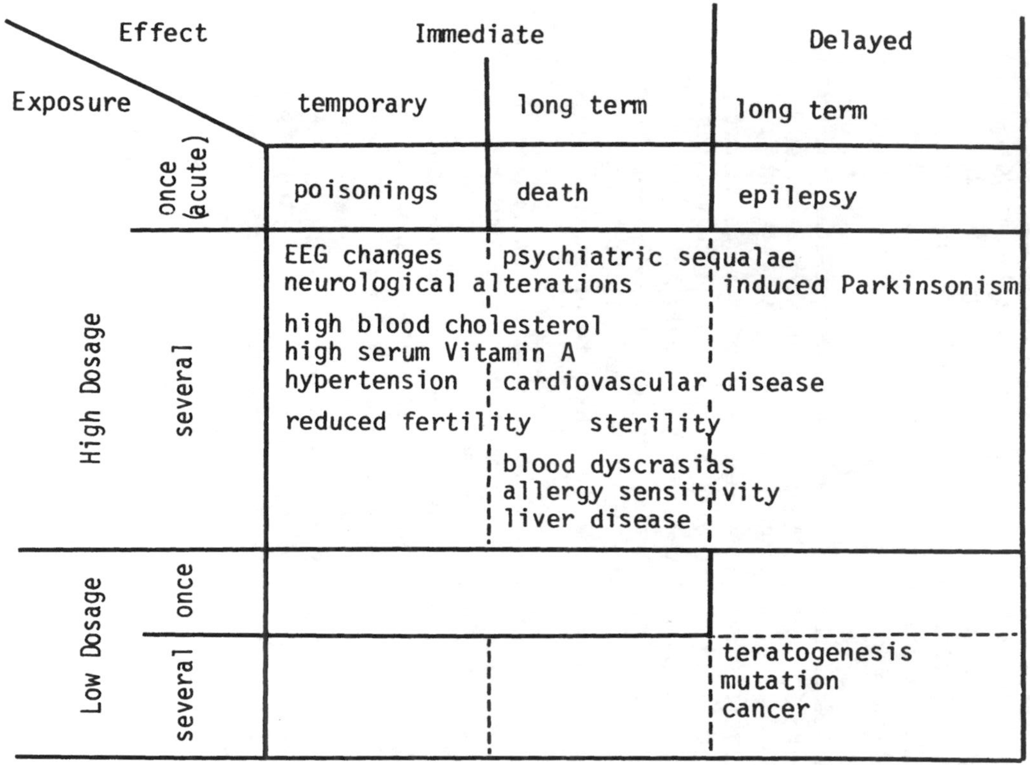

Effect	Immediate		Delayed	
Exposure	temporary	long term	long term	Delayed

High Dosage
- once (acute)
- several

Low Dosage
- once
- several

poisonings — death — epilepsy

EEG changes
neurological alterations | psychiatric sequalae
induced Parkinsonism

high blood cholesterol
high serum Vitamin A
hypertension — cardiovascular disease
reduced fertility   sterility

blood dyscrasias
allergy sensitivity
liver disease

teratogenesis
mutation
cancer

Figure 2. Some effects of pesticides on humans classified by concentration and frequency of exposure and time of onset and duration of effect. Dotted lines represent uncertainty. Starred effects are discussed in text.

are needed (Nora et al., 1967; Koos and Longo, 1976). Several pesticides are mutagenic (Epstein and Legator, 1971), but it is still unknown whether they are mutagenic in humans (Kiraly et al., 1977; Kraybill, 1977). Pesticides have also been implicated in the incidence of cancer (Cassarett et al., 1968; Radomski et al., 1968; Dacre and Jennings, 1970; Komarova, 1976; Wassermann et al., 1976, 1978).

We restrict our observations to acute exposure resulting in medically treated poisonings or death, and the estimated social cost due to pesticide-related cancer (Figure 2).

Although there is no direct epidemiological evidence that pesticides will cause cancer, the indirect evidence implicates them strongly. Pesticides cause chromosomal aberrations in human lymphocytes (Dubinin et al., 1967; Chang and Kassen, 1968; Pilinskaya, 1970; Hoopingarner and Bloomer, 1970; Czeizel et al., 1973; Yoder et al., 1973; van Bao et al., 1974; Czeizel and Kiraly, 1976; Kiraly et al., 1977, 1979). Thus they may have the potential for disrupting the normal cell cycle by mutation. Twenty-six pesticides have been found to be carcinogenic in at least one laboratory animal (Kraybill, 1977), and some may react to form carcinogens (Maugh, 1973; Wolfe et al., 1976). No one would deny that pesticides have the potential to cause cancer in humans, but whether this potential is actually realized remains to be documented. In an epidemiological study, Clark et al. (1977) reported a significant correlation between the intensity of cotton and vegetable farming and total cancer and lung cancer mortalities in southeastern United States. Other "major crops such as corn, which receive less pesticide treatment, were not significantly associated with cancer mortality." The findings of this study have important limitations as indicated by the investigators, but deserve further investigation.

The Clark et al. (1977) study reported that cotton and vegetable farming accounted for 1.6 to 6.7% of the total cancer variance in their sample. Schotterfeld (1978) estimated that the fraction of human cancer attributable to pesticides is probably less than 1%. Assuming that only 0.5% of all human cancer is due to pesticides, then with annual opportunity costs calculated to be $25 billion (OSHA, 1978) the annual cost due to pesticides is $125 million. Thus, even if the incidence of pesticide-induced cancer is low, the cost borne is fairly high. Although these data on chronic effects of pesticides have serious limitations and are extremely difficult to measure, we believe that the $125 million is a low estimate of these costs.

Medically-treated poisonings and pesticide-caused death are more easily diagnosed and more frequently reported than the effects from chronic exposure. Nevertheless, misdiagnosis and poor reporting still plague the data. With this important caveat, we report the annual number of pesticide caused deaths, hospitalized poisonings, outpatient-treated poisonings, and emergency room-treated poisonings (Table 1). The Poison Control Centers treat about 5,000 patients for pesticide poisoning each year (Lisella et al., 1975), but since the more serious poisonings are referred, we do not count them here.

The number of fatalities from pesticide poisonings has declined significantly in the past 20 years, but in 1974 there were still 52 accidental deaths (Hayes and Vaughn, 1977). The number of intentional deaths from pesticides is about three times the accidental deaths (Reich et al., 1968a; Maddy, 1978). The total estimated mortality from pesticides is about 200 per year (EPA, 1976).

Many persons who are poisoned by pesticides are rushed to hospitals. EPA (1976) estimated that an average of 2,831 of these poisonings are admitted to hospitals each year. Other data indicate that this estimate may be only one-half the real incidence of hospitalization (Cann et al., 1958; Hayes, 1960, 1964; Richardson, 1973; Lande, 1974; EPA, 1974; Maddy, 1978). Of these estimated 2,831 hospitalized poisonings, about 1000 are occupationally related (EPA, 1976). In addition to these inpatients, approximately 12,220 emergency room-treated pesticide poisonings are handled each year (CPSC, 1976).

Although many more human pesticide poisonings are treated as outpatients by private practitioners than are treated in hospitals, the numbers can only be estimated. Blondell (1978) calculates that there are 15 outpatient cases for every hospitalized case, and Hayes (1964, 1969) and West and Milby (1965) believe that there are 100 poisoning cases of all types for every fatality. These suggest 42,500 and 20,000 human poisonings, respectively. Our analysis of several studies from North and South Carolina, Florida, Texas, Pennsylvania, Oklahoma, and California (Reich et al., 1968a, b; Davis et al., 1969; Keil et al., 1970; Smith and Wiseman, 1971; Whitlock et al., 1972; Richardson, 1973; Lande, 1974; Gehlbach et al., 1974; Howitt, 1975; Caldwell and Watson, 1975; Maddy, 1978) indicated that the number of outpatients treated by private

physicians is about 30,000 per year. The proportion of
these outpatients who were occupationally poisoned is
unknown.

Thus, a total of about 45,000 medically treated human
pesticide poisonings occurs annually in the United States
(Table 1). Although the data are limited and the extrapola-
tions are tentative, a significant number of human poisonings
occur. But since a large number of poisoned workers do not
even go for medical treatment (Swartz, 1974; Howitt, 1975;
Bogden et al., 1975; Quinones et al., 1976; Owens et al.,
1978) this estimate is probably low.

To calculate the annual economic costs of human pesticide poisoning in the United States, we used two methods
(Table 1). In considering the medically treated poisonings
and costs due to cancer we calculated medical expenditures
and physician fees. If the poisoning was occupationally
related, work income lost was also added. Only accidental
deaths were cost accounted, as some would argue that the
150 odd suicides and homicides represent no costs attributable to pesticides. We have valued an individual life at
about $1 million by averaging the willingness of industry
and the government to pay for safety devices that prevent
fatality (Rhoads, 1978).

Although we believe that life and freedom from unwar-
ranted suffering cannot be accurately measured, we calculate
the annual cost of human pesticide poisonings to be about
$184 million. Since the degree of occurrence of a number
of effects is unknown, this estimate is much too low.

Domestic Animal Poisonings and Contaminated Livestock Products

Human carelessness and animal curiosity occasionally
result in poisonings of domestic animals. Some serious
poisonings of valuable animals are reported to veterinarians
for treatment, but some are not. In addition, some livestock
products become contaminated with pesticide residues and
may be destroyed or confiscated by government officials.

To estimate the number of pesticide poisoning cases
that occur in domestic animals, data from veterinary surveys
from South Carolina (Caldwell et al., 1977) and Arkansas
(Ramsay et al., 1976) were used. It is unfortunate that
the available data are from only two states; data representative of the diversity of livestock and crop production systems in the remaining states would be desirable.

Table 1. Calculated economic costs of human pesticide
poisonings and human cancer annually in the United States.

Human Poisoning Costs	Total Costs

I - Cost of Hospitalized Poisonings

2,831 hospitalized poisonings[a] x 3.7 days
in hospital[b] x \$127.70/day hospital
fee[c]   \$1,337,619

2,831 hospitalized poisonings x 3.7 days
in hospital x \$16.04/day doctor fee[d]   168,014

1,000 worker hospitalized poisonings[e] x
6.67 days lost work[f] x \$34/day[g]   226,780

II - Cost of Nonhospitalized Poisonings

30,000 physician treated x 1.5 physician
visits[h] x \$20/visit[i]   900,000

40% nonhospitalized physician treated[j]
x 42,200[k] physician treated x 6.67
days lost work x \$34/day[l]   3,828,046

III - Cost of Emergency Room Treated Poisonings

12,200 Emergency Room poisonings x
\$25/visit[m]   305,000

IV - Cost of Fatalities

52 Accidental Fatalities[n] x \$1 million[o]   52,000,000

V - Cost of Human Cancer Due to Pesticides

0.5% cancer[p] x \$25 billion[q]   \$125,000,000

| | TOTAL | \$183,765,459 |

Footnotes to Table 1. (facing page)

a/ EPA, 1976.
b/ Average 3.7-day stay in the hospital for pesticide poisoning (Daniel-Guido, 1978).
c/ Hospital cost/day exclusive of doctor fees (HII, 1976).
d/ Average cost of general practitioner's or internist's visit in the hospital (AMA, 1977).
e/ Estimated from EPA, 1976.
f/ Average number of days of work lost per pesticide incident (State of California, 1974).
g/ Wage computed by averaging wage of agrichemical workers with that of farmers and agricultural workers (USDL, 1975; USDA, 1977).
h/ Assume each poisoning victim visits a medical doctor 1.5 times.
i/ Fee per visit including medication (AMA, 1977).
j/ Assume 40% of nonhospitalized physician-treated cases were employed adults. Estimated from EPA (1976), which states 39% of hospitalized poisonings were children under 4 years old and Lisella et al. (1975) who states 68% of all poisonings were children.
k/ 30,000 physician-treated poisonings + 12,200 emergency room-treated poisonings = 42,200.
l/ Overall worker average daily wage (USDA, 1977).
m/ Lisella et al., 1975.
n/ A total of 52 accidental deaths from pesticides of a total of 217 pesticide poisoning fatalities.
o/ Estimated value of human life is assumed to be $1 million.
p/ Assumed incidence of cancer due to pesticides.
q/ OSHA, 1978.

Table 2. Animal pesticide poisoning cases calculated for the United States.

Species	Number in U.S. (x 10^3)	Percentage of Pesticide Poisonings [d]	Number of Pesticide Poisoning Cases (x 10^3)	Vet. Costs [e] (x 10^3)	Percentage of Fatal Pesticide Poisonings [f]	Number of Fatal Pesticide Cases (x 10^3)	Cost of Fatalities (x 10^3)	Total Losses (x 10^3)
Cattle	128[a]	0.0144	18.4	$552	0.007	9.0	$2,250[g]	$2,802
Dogs	41[b]	0.2180	74.1	2,223	0.050	17.0	850[h]	3,073
Horses	50[c]	0.0143	7.2	216	0.007	3.5	1,400[i]	1,616
Cats	23[b]	0.0478	14.8	444	0.035	10.9	55[j]	499
Swine	39[a]	0.0037	1.4	42	0.002	0.8	562[k]	604
Poultry	1,300	0.0001	1.3	7	0.0001	1.3	1	8
TOTAL								$8,602

[a] USDA, 1976b.
[b] Anonymous, 1978. (Note, total dogs and cats, both tame and wild, is about 100-120 million (Wittwer, 1975)).
[c] Estimated.
[d] Percentages based on incidence of cases in Arkansas and South Carolina (Ramsay et al., 1976; Caldwell et al., 1977).
[e] Calculated based on $30 per incident, except for poultry.
[f] Percentages based on incidence of cases in South Carolina (Caldwell et al., 1977).
[g] Valued at $250/head (USDA, 1976b).
[h] Estimated value at $50/dog (no attempt was made to attach a personal or social value to pet dogs).
[i] Estimated at $400/horse.
[j] Estimated value at $5/cat (no attempt was made to attach a personal or social value to pet cats).
[k] Valued at $72/pig (USDA, 1976b).

We found the highest incidence of poisonings occurs in cats and dogs (Table 2). These animals probably have a greater opportunity to contact pesticides than other domesticated animals because they wander freely about the home and farm, and therefore are poisoned more frequently. Veterinary costs for the treatment of dog poisonings and the cost of fatalities to horses and cattle account for about 70% of the cost of pesticide poisonings in all domestic animals. In other words, the valuable or frequently exposed animals account for the highest costs in direct animal poisonings. We calculated that $8.6 million a year are lost from direct poisonings of domestic animals (Table 2). Since this estimate is based only on poisonings reported to veterinarians, it is probably low. When a poisoning occurs and little can be done for the animal, the farmer seldom calls a veterinarian (Maylin, 1977). Also, mild cases are seldom reported.

Some of these costs may be internalized into market pricing structures, but it is not clear whether this type of loss is actually internal.

Sublethal exposure may change the quality of livestock and livestock products, but little is known about this. Regardless, such sublethal exposure may lead to meat and milk contamination with pesticide residues. If residue levels exceed a given threshold concentration, the products are deemed a public health hazard and confiscated by government officials.

Approximately 1% of the livestock entering state and federally inspected slaughter houses is inspected for pesticide residues (Clark, 1977). The value of this condemned meat is $3.1 million (Clark, 1977). This estimate does not include the losses from delaying slaughter until pesticide residues decline to acceptable levels. Since only 1% of the livestock is inspected, some pesticide-contaminated meats may evade notice and would serve to increase dietary exposure in human populations.

Milk is also inspected for pesticide residues and contaminated milk is disposed of and not used. When contamination occurs because of an accident or for reasons not under the control of the farmer, the U.S. government will compensate the loss. In 1977, about $143,500 was paid in compensation under the U.S. milk indemnity act. This probably represents about two-thirds of the total losses due to pesticide residues in milk (Schiermeyer, 1977). When it is contaminated because of the farmer's carelessness, or any other unindemnifiable source, there is no public

record of the loss. Thus, the total monetary loss due to
the pesticide contamination of milk is about $210,000 per
year. Whether these costs are internalized is not clear,
that they represent additional social and environmental
costs is a fact.

The combined costs from domestic animal poisonings and
contaminated livestock products amount to at least
$11,910,000 per year.

Increased Pest Control Costs

Every crop has at least one major pest insect, but the
vast majority of insects and mites are only minor pests or
unimportant because natural enemies control these potential
pest populations at subeconomic levels. However, when
insecticides or other pesticides are applied to control one
pest, the natural enemies of another potential pest are
sometimes inadvertently killed. This may result in secondary
pest outbreaks, necessitating the use of added treatment to
control the new pest population.

In addition to this cost due to the loss of natural
enemies, the widespread use of pesticides and extensive
exposure of pest populations has often successfully selected
for pesticide resistance. Additional applications of the
commonly used insecticides and/or a substitution of a more
expensive pesticide is frequently used to control these
resistant populations. Since 364 insect and mite species
are known to be resistant to at least one pesticide (Georghiou
and Taylor, 1977), the increasing levels of pesticide
resistance in pest populations incur large environmental
costs -- costs initiated in the past but paid for now.

To estimate these costs, we surveyed the literature
for known cases of secondary pest outbreaks and evolving
resistance to determine the proportion of application costs
due to these causes. Then, by a modified Delphi technique,
several entomologists[1] in various parts of the country were

---

[1] Perry Adkisson, Texas A&M University; Max J. Bass, Auburn
University; Brian Croft, Michigan State University; Charles
J. Eckenrode, N.Y.S. Agricultural Experiment Station,
Geneva, New York; George Georghiou, University of California,
Riverside; E. H. Glass, N.Y.S. Agricultural Experiment
Station, Geneva, New York; Carl B. Huffaker, University of
California, Riverside; William Luckmann, University of
Illinois; L. Dale Newsom, Louisiana State University;
Robert Rabb, North Carolina State University, Thomas E.
Reagan, North Carolina State University; George Teetes,

asked to validate our estimates and the data were revised
based on their advice. This process was repeated 3 times.
The data are presented in Table 3 for the 38 crops we
explicitly studied.

Cotton and corn account for the majority of the costs,
but as 64% of all agricultural pesticide used in the United
States is applied to these two crops, this is not surprising.
Large costs also occur on sorghum, apples, and potatoes.

Cotton is grown in three areas of the country, each
with its own arthropod pest control problems. The Southeast,
plus a portion of Texas, is dominated by the boll weevil;
the key pest of northern Texas and Oklahoma is the pink
bollworm; and in the irrigated West, which includes Arizona,
the primary pests are the lygus bug and the pink bollworm
(Adkisson, 1973; Frisbie and Walker, 1979). Most of the
other pests have become serious problems secondarily since
the heavy use of insecticides was begun (Newsom, 1962;
Adkisson, 1973; Stern, 1976; Bottrell and Rummel, 1978),
including the widespread cotton budworm and cotton bollworm
(Wille, 1951; Brazzel et al., 1953; Newsom, 1962; Ridgway
et al., 1967; Laster and Brazzel, 1968; Lingren et al.,
1968; Ridgway and Lingren, 1972; Cate et al., 1972; Lingren
et al., 1972; Pate et al., 1972; Van Steenwyck et al.,
1975; Johnson et al., 1976a; Adkisson, 1977; Plapp and
Vinson, 1977; Kinzer et al., 1977; Pimentel et al., 1977a),
the cotton aphid (Bartlett, 1968), the beet armyworm and
the cabbage looper (Newsom, 1962; Falcon et al., 1968;
1971; Ehler, 1972; Eveleens, 1972; Ehler et al., 1973;
Eveleens et al., 1973; Gutierrez et al., 1975; Ehler,
1977; Ehler and Miller, 1978), and spider mites (Newsom,
1962; Bartlett, 1968).

Twenty-five cotton pests have evolved resistance to a
number of pesticides, including the cotton bollworm, the
cotton budworm, the boll weevil, the pink bollworm, and the
lygus bug (Taylor and Headley, 1975).

The situation that developed in northeastern Mexico
and the lower Rio Grande in Texas is a striking example of
the magnitude of environmental and social costs that can be
incurred by the evolution of resistance. Because the
tobacco budworm, a pest on cotton, evolved a high level of
resistance to four major classes of insecticides, in early

Texas A&M University; Ward M. Tingey, Cornell University;
Robert van den Bosch, University of California, Berkeley;
William Whitcomb, University of Florida.

Table 3. Estimates of the environmental costs due to reduction in natural enemy populations and insecticide resistance.

Crop	Acres[a] x 10³	% Acres Treated[b]	Insecticide Treatment Costs $/Acre[c]	Total Insecticide Control Cost $ x 10³	% Cost of Treatments		Total Added Insecticide Cost	
					Due to loss of natural enemies[d]	Due to increased insecticide resistance[d]	Due to loss of natural enemies $ x 10³	Due to increased insecticide resistance $ x 10³
Corn	65,194	52	7	237,306	1	25	2,373	59,327
Cotton	12,547	95	20	238,393	40	15	119,197	35,759
Wheat	65,459	7	8	36,657	0	0	0	0
Soybeans	52,460	8	8	33,574	5	0	1,679	0
Rice	2,569	35	8	7,193	10	0	719	0
Tobacco	963	77	20	14,830	5	5	741	741
Peanuts	1,472	87	16	20,490	10	0	2,049	0
Sorghum	13,917	39	6	32,566	15	15	4,885	4,885

Table 3. (continued)

Sugar Beets	1,217	30	14	5,111	10	5	511	255
Other Grain	38,000	3	8	9,120	0	0	0	0
Alfalfa	26,642	8	8	17,051	5	5	853	853
Other Hay	33,904	0.5	8	1,356	0	0	0	0
Other Field Crops	6,533	13	8	6,794	0	0	0	0
Pasture	563,000	0	0	0	0	0	0	0
Total	883,877			660,441			133,007	101,820

a/ USDA, 1975b.

b/ Corn and cotton data were obtained from a survey by Pimentel et al. (1977a) and all others are from USBC, 1973a; USDA, 1975a.

c/ USDA, 1975c, treatment costs in this publication were doubled to arrive at treatment cost per acre.

d/ Estimated, see text page 14.

Table 3. (continued)

Crop	Acres[a] x 10³	% Acres Treated[b]	Insecticide Treatment Costs $/Acre[c]	Total Insecticide Control Cost $ x 10³	% Cost of Treatments		Total Added Insecticide Cost	
					Due to loss of natural enemies[d]	Due to increased insecticide resistance[d]	Due to loss of natural enemies $ x 10³	Due to increased insecticide resistance $ x 10³
Lettuce	226	89	50	10,057	10	5	1,006	503
Cole	196	90	50	8,820	10	5	882	441
Carrots	80	57	8	365	0	0	0	0
Potatoes	1,380	77	16	17,002	15	20	2,550	3,400
Tomatoes	465	93	32	13,838	5	10	692	1,384
Sweet corn	628	70	16	7,034	10	10	703	703
Onions	110	80	32	2,816	5	25	141	704
Cucumbers	177	59	16	1,671	0	5	0	84
Beans	444	70	15	4,662	5	5	233	233

Table 3. (continued)

Cantaloupe	70	86	15	903	0	5	0	45
Peas	426	56	15	3,578	0	5	0	179
Peppers	48	77	15	554	5	10	28	55
Sweet								
Potatoes	119	67	15	1,196	0	5	0	60
Watermelons	215	62	15	2,000	0	5	0	100
Asparagus	112	47	15	790	0	0	0	0
Other								
Vegetables	128	70	15	768	0	5	0	67
Total	4,824			76,054			6,235	7,958

a/ USDA, 1975b.

b/ USBC, 1973b.

c/ USDA, 1975c, treatment costs in this publication were doubled to arrive at treatment cost per acre.

d/ Estimated, see text page 14.

Table 3. (continued)

Crop	Acres x 10³	% Acres Treated[f]	Insecticide Treatment Costs $/Acre[g]	Total Insecticide Control Cost $ x 10³	% Cost of Treatments Due to loss of natural enemies[h]	% Cost of Treatments Due to increased insecticide resistance[h]	Total Added Insecticide Cost Due to loss of natural enemies $ x 10³	Total Added Insecticide Cost Due to increased insecticide resistance $ x 10³
Apples	526[a]	91	54	25,848	20	10	5,170	2,585
Cherries	129[a]	66	54	4,598	5	5	230	230
Peaches	301[a]	76	40	9,150	10	10	915	915
Pears	112[a]	46	40	2,061	5	30	103	618
Prunes & Plums	161[a]	72	40	4,637	15	10	696	464
Grapes	720[b]	67	40	19,296	10	10	1,930	1,930
Oranges	862[c]	88	25	18,964	20	5	3,793	948
Grapefruit	174[c]	81	20	2,819	20	5	564	141

Table 3. (continued)

Lemons	83[c/]	82	20	1,361	20	5	272	68
Other Citrus	87[d/]	22	20	383	20	5	77	19
Strawberries	40[e/]	70	10	280	5	10	14	28
Other Fruits	147[d/]	44	10	647	20	5	129	32
Pecans	382[a/]	60	14	3,209	0	5	0	160
Walnuts	415[e/]	60	14	3,486	10	5	349	174
Total	4,139			96,739			14,242	8,312

a/ USBC, 1973c.

b/ USDA, 1975d.

c/ FA, 1975.

d/ USDA, 1975b; FA, 1975; CCLRS, 1975.

e/ USDA, 1975b.

f/ USBC, 1973c; USDA, 1975a.

g/ USDA, 1975c, treatment costs in this publication were doubled to arrive at treatment costs per acre.

h/ Estimated, see text page 14.

1970 approximately 700,000 acres of cotton, with a market
value of about $135,000,000, had to be abandoned (Adkisson,
1971, 1972; NAS, 1975). The economic and social impact on
the farming communities that depended on cotton was devasta-
ting. But the budworm is not the only pest with DDT,
cyclodiene, organophosphate, and carbamate resistance.
Others include an ixodid mite, the Colorado potato beetle,
the alfalfa weevil, the rice weevil, the corn weevil,
blowfly, two Anopheles mosquitoes, the common housefly, a
nitidulid Meligethes aeneus, the green peach aphid, the
cotton leaf perforator, the corn earworm or cotton bollworm
or tomato fruitworm, the beet armyworm, S. littoralis, the
cabbage looper, the diamondback moth, P. xylostella, and
the German cockroach (Georghiou and Taylor, 1977). The
crop failure on the Rio Grande may not be a unique incident.
The production losses and disastrous local social consequences from this boom and bust cycle brought on by the
collapse of chemical control systems on crops are not
included in Table 3.

Corn is the most valuable crop grown in the United
States. Covering over 71 million acres in 1976 (Luckmann,
1978), it is susceptible to pest attack. Protection relies
on pesticides, though cultural methods and resistant culti-
vars are also important. Major pests include the European
corn borer, the black cutworm, the fall armyworm, the corn
leaf aphid, the northern corn rootworm, the western corn
rootworm and the southern corn rootworm. In 1961 the
western corn rootworm evolved resistance to cyclodienes.
This resistance became widespread in 5 years time. Although
the substitute insecticides did not provide the same level
of control as aldrin and dieldrin initially did (Reynolds,
1977), the rootworms are now also resistant to them (Luckmann,
1978).

Because of high cosmetic standards on most orchard
fruits, apple orchards are heavily sprayed. They receive
as many as 30 treatments a year for control of the apple
maggot fly, the plum curculio, the codling moth, the red-
banded leaf roller, and other minor pests, depending on the
area of the country (Croft, 1978). These sprays kill the
coccinellid and phytoseiid mite predators of the San Jose
scale, the oyster-shell scale, the apple aphid, the rosy
apple aphid, the woolly apple aphid, the white apple leafhopper, the potato leafhopper, the European red mite, the two-
spotted spider mite, and the apple rust mite, causing outbreaks of these secondary pests (Croft, 1978). The high
levels of pesticide exposure have selected resistance in
several pests including the apple maggot fly, and mites

(Georghiou and Taylor, 1977; Croft, 1978), accentuating the control problems of producing marketable fruit.

Sorghum, consumed directly as food and used as fodder, is one of the more important crops grown in the United States. Its key pests are the sorghum midge, the sorghum stem borer, the shoot fly, and the greenbug (Young and Teetes, 1977). Relatively small amounts of insecticide are applied to sorghum and this is used mainly to control the greenbug. Greenbug control with insecticides reduces natural enemy populations (Ward et al., 1970), allowing spider mites to outbreak (Young and Teetes, 1977), and has selected resistant greenbug populations (Teetes et al., 1975; Peters et al., 1975).

Pesticides also interact in other ways that result in pest outbreaks. The use of fungicides may contribute to pest problems by reducing populations of entomogenous fungi. The application of benomyl, toxic to these fungi, results in increased survival of velvet bean caterpillars and cabbage loopers in soybeans and eventually leads to reduced crop yields (Ignoffo et al., 1975; Johnson et al., 1976b). Pesticides may change the soil microfauna composition. Application of Furadan to soil probably alters the microflora, resulting in more rapid biological degradation of carbamate insecticides (Williams et al., 1976), which would reduce their effectiveness on soil insects like the corn rootworm complex.

Although livestock pests, disease vectors, glasshouse pests, and other nonagricultural insect pests have natural enemies, we have assumed that the loss of natural enemies in this case is not significant. Little study has been done on this problem. On the other hand, numerous pests have become resistant to pesticides (Georghiou and Taylor, 1977). Since a relatively small quantity of pesticide is used for control of noncrop pests, we calculated that resistance in insect and mite pests of livestock and man amounts to only $15 million annually.

The total increased control cost that results from the destruction of natural enemies and pesticide resistance amounts to at least $287 million annually (Table 3). This figure, which we recognize as an indirect cost, is reflected by the increased consumption of pesticides and is already accounted for in the $2.2 billion market costs of pesticides. The total indirect costs of local crop failure from a secondary pest outbreak or uncontrollable pesticide resistance have not been estimated and are not included here.

Neither are losses due to plant pathogens resistant to
fungicides and bactericides (Ogawa et al., 1976).

Honey Bee Poisoning and Reduced Pollination

A great deal of pesticide is directed against insects,
so it is not surprising that they are often toxic to honey
bees and other insect pollinators. These reduced pollinator
populations cause significant indirect costs related to
loss of honey and crop production. The effects of the loss
of wild pollinators on the natural ecosystem are not well
known, so they represent unknown indirect costs.

The impact of widespread pesticide use has been particularly adverse to the beekeeper, killing hives, reducing
hive viability and honey production, increasing maintenance
costs, and decreasing the available pollen and nectar
supplies.

Recognition of these problems has resulted in the
legislation of the Bee Indemnity Act of 1970 to compensate
apiculturalists for these losses (Public Law 91-524).
Since 1970 the Agricultural Stabilization and Conservation
Service has paid a total of about $21 million in bee indem-
nities to apiculturalists for their losses (ASCS, 1976).
This probably represents only a small portion of actual
losses.

Martin (1978) estimated that perhaps 20% of all honey
bee colonies are actually affected by pesticides, including
approximately 5% of the colonies that are killed outright.
Other colonies may die during the winter because they were
weakened by pesticides or suffered losses when apiculturalists moved them to avoid pesticide damage. The loss from
colony kills and partial kills, reduced honey production,
and movement of colonies totals about $21.1 million annually
(Martin, 1978). Also, as a result of heavy pesticide use
on certain crops, beekeepers are excluded from 10 to 15
million acres of good apiary location (Martin, 1978). The
estimated loss in honey production from these regions is
$22.5 million annually (Martin, 1977).

The pollination and yield of many fruits, vegetables,
and forage crops depend heavily on honey bees and wild bees
(McGregor, 1976). Supplemental pollination by honey bees
is essential for the economic production of some crops.
For example, apples and almonds produce almost no crop, and
alfalfa seed yields are nil unless they are insect pollinated
(McGregor, 1976). In these crops, in order to minimize
hazard to bees, growers are usually careful to avoid spraying

during the bloom period. However, accidents occur, resulting in losses to the grower and beekeeper and in other losses associated with beekeeper's fear of damage to their hives. In 1977 the blueberry crop in New Jersey was small, partly because a cold spring set back the prepollination spray and dust schedules. The fear of loss to the delayed sprays and dust made the beekeepers reluctant to move in their hives when the blueberries bloomed. As a result many early varieties were not adequately pollinated and yields were reduced (Stricker, 1977). The total reduction of yield from these accidental poisonings of honey bees and the threat of accidental poisonings is largely unknown.

Other crops are self-fertile and produce a substantial crop even in the absence of pollinators. Nevertheless, bees have been found to enhance yields in some cases, such as cotton, soybeans, flax, and some varieties of citrus (McGregor, 1976). For example, both crop caging tests (Shishikin, 1946; Mahadevan and Chandy, 1959) and pollinator supplementation tests (McGregor et al., 1955) have shown for several cotton varieties that good pollination by bees can result in yield increases of 20 to 30%. Good pollination on cotton, however, has not been possible because the intensive use of insecticides excludes pollinators on cotton (McGregor, 1976). If cotton yield increased only 10% after efficient pollination, and subtracting the charges for honey bee rental necessary to accomplish this, the net annual gain could be as high as $300 million.

Atkins (1977) emphasizes that poor pollination will reduce crop yields and also points out that it will reduce the quality of crops such as melons and various fruits. He reported that with adequate pollination, melon yields were increased 10%, but quality was increased 25% as measured by dollar value of the crop.

Estimates of annual agricultural losses from the poor pollination of crops by honey bees due to pesticides range from about $80 million (Atkins, 1977) to a high of $4 billion (McGregor, 1977). The conservative estimate of $80 million represents our smallest estimate of these production losses.

Although all the effects of pesticides on the economically important honey bee-crop interaction are difficult to measure, the effects on other pollinators are even less easy to quantify. Wild pollinators are killed in pesticide-treated forests (Kevan, 1975; Plowright et al., 1978). The use of herbicides on crops, roughlands, and wastelands reduces the amount of time during the year that pollen and nectar are available to bees by reducing the diversity of flowering plants. This food limitation reduces wild bee populations (Levin, 1970; McGregor, 1973). The result is that farmers have to rent increased numbers of commercial honey bee colonies to provide adequate pollination of their crops. In California, about 700,000 colonies of honey bees must be rented annually at $8 per colony to supplement natural pollination of almonds, alfalfa, melons, and other fruits and vegetables produced for seeds (Atkins, 1977). Since California produces nearly 50% of our bee-pollinated crops (calculated in terms of dollar values), the total cost for bee rental is about $11.2 million for the entire country. Since much of this rental is needed because of our extensive crop monoculture system, only one-tenth (about $1 million) is considered due to the loss of wild pollinators.

Little is known about the effects of wild pollinators on the natural ecosystem. Bumble bee losses from fenitrothion resulted in reduced seed set in red clover (Plowright et al., 1978), but whether the ramifications significantly affect ecosystem structure or function is unknown.

The total calculable environmental and social costs from reduced pollination and honey bee losses totals about $135 million per year. Clearly the available evidence confirms that annual honey bee losses, and agricultural losses from poor pollination due to honey bee and wild bee kills are significant and the problems deserve careful scrutiny.

Persistent pesticides such as many organochlorides could be applied to crops without much hazard to bees. Modern, less persistent materials provide a much greater danger to bees, both because they are often extremely toxic to hymenoptera and because their decreased persistence often requires insecticide sprays during the crop bloom periods (Johansen, 1977). Direct losses of honey bees and indirect losses due to incomplete pollination seem therefore to be problems that are accelerating.

Crop and Crop Product Losses

Pesticides applied to protect crops sometimes damage those crops and other valuable plants near the site of application. Although there is at least one report of fungicide applications indirectly reducing crop growth (Dubey, 1970), herbicides are generally responsible for this type of damage. There are several ways economic plant loss can occur. The treated crop may be damaged if pesticides are applied improperly or under unfavorable environmental conditions. When excessive pesticide residues accumulate on the crop, harvested products are often devalued or destroyed. Other crops can be damaged when pesticide drifts onto them from a nearby treated crop or when herbicide residues accumulate to toxic levels in the soil, preventing chemical-sensitive crops from being planted in rotation or inhibiting the growth of crops that are planted. In addition, the widespread use of herbicides has caused changes in common weed populations, promoting the growth of weeds that are difficult to control by conventional means.

Under ideal conditions, application of recommended pesticide dosages has minimal effect upon that year's crop yields (Chang, 1965; Elliot et al., 1975). If, however, weather or soil conditions are unsatisfactory, standard herbicide treatments may cause yield reductions ranging from 2 to 50% (von Rumker and Horay, 1974; Elliot et al., 1975; Akins et al., 1976). Improper pesticide application procedures such as the use of less desirable pesticides, poor or ill-timed application techniques or application at incorrect dosage rates annually result in significant losses (Hahn, 1977; Duke, 1977). Even the most careful herbicide applications can result in scattered bare spots if the applicator must slow or stop the spraying rig in the field or accidentally overlaps areas of the field (Sweet, 1977). Furthermore, as demonstrated with corn, herbicides can increase the susceptibility of the crop to insects and diseases (Oka and Pimentel, 1974; Pimentel et al., 1978b).

An additional loss is incurred when food crops are seized for exceeding the FDA regulatory tolerances for pesticide residue concentrations. This may occur because of poorly timed application schedules, but more frequently results from delayed residue breakdown or forced early harvests. This loss is estimated to be at least $2.5 million (Pimentel et al., 1977b).

Because of the persistence of some widely used herbicides in the soil, future crops may suffer damage from previous years' use. Crops planted in rotation that are herbicide sensitive may be injured, and opportunity costs from restricted rotation options or forced continuous planting can be incurred.

Rotation crop losses are especially pronounced when exceptionally cold or dry spring weather inhibits herbicide decomposition. For example, in rotations from corn to small grains or beans in New York, while some damage appears in an average of 3% of the small grain fields, spring weather conditions in 1977 did not favor herbicide decomposition, so approximately 10% of the fields reported damage (Hahn, 1977). Small grain and bean yields were reduced by an estimated 1% due to this carry-over damage. Indiana also reported exceptionally severe damage that year (Bauman, 1978). The significant feature of this type of carry-over problem is its unpredictability. This characteristic makes it difficult to estimate losses.

Persistence may decrease crop rotation choice or force continuous culture. Continuous planting of some crops may result in intensified insect, weed, and pathogen problems (PSAC, 1965; NAS, 1975; Pimentel, 1977). These problems will reduce yields and/or require increased investments in pesticides. The costs of restricted crop rotation choice are difficult to estimate (Slife, 1972).

Drifting pesticides often cause significant crop loss. Although drift occurs with any application method, it is most pronounced with aircraft application. Typically, 20% to 80% of the pesticides applied by air misses the target area (Yates and Akesson, 1973) and may injure nontarget crops 20 miles downwind (Henderson, 1968; Akesson et al., 1978).

About 65% of all agricultural pesticides are applied by air, so this is a potentially serious problem (USDA, 1976a). In addition to the agricultural pesticides, about 17 million pounds of herbicides are used annually by highway and utility crews to clear roadsides and rights-of-way (NAS, 1975). Drift damage to field crops, gardens, tree crops, and shelter belts arising from these applications is reported yearly (Elmore, 1977; Knake, 1977).

Herbicide drift, particularly in regions with diverse cropping patterns, commonly injures nontarget crops. For example, drift injury to the Washington grape crop has been "a very serious problem" (Fox, 1978) since the mid-1950s, and individual growers may suffer losses of over 50% because of it. In fact, because grapes are so sensitive to some common herbicides, their use has been restricted in the vicinity of some vineyards (Akesson et al., 1978). Drift injury has been reported in Oregon on sugar beets, potatoes, fruit crops, and almonds (Brown, 1978a), in Indiana on tomatoes and soybeans (Bauman, 1978), in Mississippi on cotton (Hurst, 1978), in California on spinach, lettuce, and pears (Elmore, 1977), in Alabama on forages (Walker, 1977), in South Dakota on soybeans (Auch and Arnold, 1978), and in North and South Dakota on sunflowers (Arnold, 1979).

Clearly, drift injury is not restricted to certain areas of the country or to a small number of crops, although some areas and some crops suffer more injuries than others. Although drift damage to crops frequently occurs, even rough data are lacking, making it difficult to calculate these losses.

The accurate estimation of crop losses attributable to pesticide use is extremely difficult. Government agencies do not keep systematic records of these losses -- indeed, because of the diffuse and erratic nature of the problem such documentation could be impossible -- and are only notified of the most severe incidents. State Cooperative Extension scientists, who hear on an average 50 cases per state per year, learn of only "the tip of the iceberg" (Bauman, 1978). Virtually no field research has been devoted to this question. Because this damage is heavily dependent upon climatic and human "accidents", extrapolation from year to year or state to state is flawed. For these reasons, accurate data do not exist, and accurate estimates are not available.

An eminent weed scientist has suggested that losses to herbicide drift and persistence in Illinois amount to from 0.1% to 0.25% of the state's annual production (Knake, 1977). We have extrapolated the lower figure of 0.1% to give an idea of the magnitude of the problem; we estimated that annual national crop losses from pesticide, drift, and persistence is on the order of $60 million. The Illinois figure is strictly applicable only to cropping patterns in the Corn Belt. In other areas of the country, cropping patterns differ, so costs are different. In the Wheat Belt, drift problems are comparatively less, while in the Southeast, Northeast and Pacific coastal states, they appear to be more severe. Pesticide persistence problems can be as serious in the Northeast and Southwest as in Illinois, while in the Wheat Belt, Northwest and Southeast, they are probably not as large. Thus, our estimate is probably low.

Many of these problems are inherent in the pesticide application process, and liability often falls on the commercial applicators. For instance, applicators sometimes are charged for damage inflicted during or after treatment, and in many states an applicator must show evidence of financial responsibility before spraying. Many applicators carry crop liability insurance to protect themselves from expensive lawsuits. Because of damage suits, annual insurance rates are now a minimum of $382 for ground applicators and

$1,982 per aircraft for aerial applicators (Turner, 1978).
Nationwide the total investment for aerial crop liability
insurance is $7.9 million (based on the estimate that 1/2
of the aircraft applying pesticides carry such insurance)
(Higbee, 1978). Although these are indirect costs, they
are probably not external costs.

A further troubling, and perhaps expensive, aspect of
herbicide use is the change in typical weed populations
that it appears to be promoting. Although weeds have the
capacity to evolve resistance to herbicides (Grignac,
1978), this is not yet a problem in growers' fields.
Instead, the species composition of weed populations is
changing. Species of weeds tolerant to widely-used herbicides are rapidly replacing intolerant species in fields
(Day, 1978). Commonly, perennial weeds are replacing
annual weeds; the perennials are generally more vigorous
and difficult to control by either chemical or mechanical
methods than the annuals they supplant. As a result, both
expenditures for weed control and losses to weeds may be
increasing.

Extensive herbicide treatments to range and forest
lands cause substantial plant community changes. The
ramifications up the food chain of any shifts in plant
populations are unknown. Changes in the plant community
will result in insect herbivore population changes. These
changes may affect the natural enemy and pollinator populations.

The total estimable costs of losses to crops and crop
products is about $70 million (Table 4).

Table 4. Estimated loss of crops and trees due to the use of
pesticides.

Crops or Trees Lost	Total Costs
Crops injured through the direct use of pesticides (0.1%)	$60,000,000
Crop Applicator Insurance	7,900,000
Crops seized as exceeding pesticide tolerances	2,500,000
TOTAL	$70,400,000

Fishery and Wildlife Losses

Pesticides and their breakdown products that run off
treated lands generally enter nearby aquatic ecosystems.
Soluble pesticides are easily washed into streams and
lakes, while others are carried with soil sediments into

water bodies. With many row crops, such as cotton and corn, water erosion carries an average of 20 tons of soil (plus pesticide residues) per acre per year into aquatic habitats (Pimentel et al., 1976).

Somewhat encouraging, however, is the fact that three successive studies have shown steadily decreasing concentrations of pesticides in surface waters and streams during the years 1964-1978 (Lichtenberg et al., 1970; Schulze et al., 1973; NYS DEC, 1977-78). This is apparently due to the replacement of the persistent pesticides with less persistent materials.

There are several ways in which pesticides are known to cause fishery losses. These include: high pesticide concentrations in water which directly cause fish kills, low level doses which may kill the more susceptible fish fry and stressed fish or eliminate essential fish foods such as insects and other invertebrates. In addition, because there are safety restrictions placed on the catching or sale of fish contaminated with pesticide residues, these unmarketable fish must be considered a loss.

Reported losses from direct fish kills have increased substantially since 1966 (Figure 3) (HEW, 1960-66; FWPCA, 1967-70; EPA, 1972-76). During the early 1960s the yearly average kills were in the range of 200,000-400,000 fish. For the last five years the average has been well over 1 million each year. These estimates of fish killed are considered to be low for many reasons. For instance, 20% of the reported fish kills give no estimate of the number of fish killed and fish kills often cannot be investigated quickly enough to determine the primary cause. Fast-moving waters wash away poisoned fish while other poisoned fish sink to the bottom and cannot be counted. Perhaps most important is the fact that, unlike direct kills, few, if any, of the widespread, low-level pesticide poisonings are observed in dramatic fashion and therefore are not recognized and reported (HEW, 1960-66; FWPCA, 1967-70; EPA, 1972-76). The recent rise in the reported fish kill may well be due to improved reporting procedures and/or to more toxic pesticides being used in our environment.

In 1978 an average value of about 40¢ per fish killed was determined (Lopinot, 1971; Sherry, 1971; AFS, 1975; ILL DEC, 1976). At this cost, which may be low, the value of the estimated 2 million fish killed per year is $800,000. This is certainly a low estimate, with the actual loss probably several times this amount.

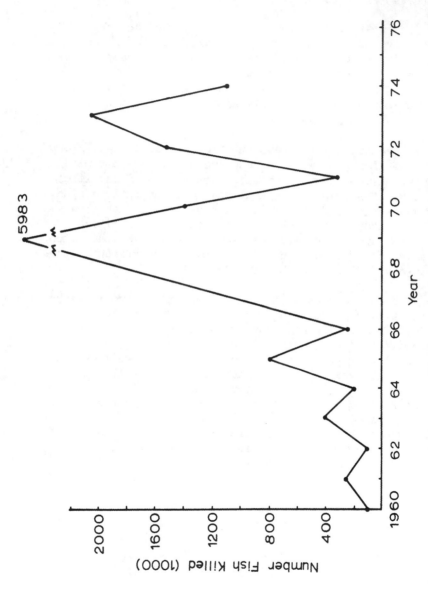

Figure 3. The number of fish killed annually in the United States (HEW, 1960–66; FWPCA, 1967–70; EPA, 1972–76).

Historically there have been periodic major industrial mishaps producing widespread pesticide contamination and resulting in massive fish kills. Examples of this type of incident have been: the large-scale endrin spill in the Mississippi in the early 1960s, a DDT spill in Los Angeles in 1969-70 (Ehrlich et al., 1977), the mirex contamination of Lake Ontario, and the Kepone contamination of the James River in 1975.

These repeated events have massive one-time costs as well as widespread longer-term costs from each event. The average yearly one-time costs, based on the past occurrence rate, as well as the continuing annual costs from recent events, are all indirect costs of pesticides. However, the extensive costs of these occurrences are almost impossible to calculate.

In the James River Kepone incident, where the extent and seriousness have been well documented, the precise dollar costs are uncertain. The lower James River was closed to fishery activity in December 1975 and will remain closed to some fisheries at least through December 1979 (it can be closed only on a year-to-year basis). Due to this, continuing fishery losses were reported to be at least $2.7 million annually (U.S. Senate, 1976). Of this, $1.1 million was from seed oyster production. The cost is actually even more extensive because the James River, as Virginia's principal source of seed oysters, cannot be replaced. The loss attributed to recreational use was reported to far exceed these commercial losses. Other one-time costs related to the Kepone incident include a fine levied against Allied Chemical of $5.3 million plus the cost to Allied of an $8 million trust fund used to establish the Virginia Environmental Endowment (Gilley, 1978). Still to be assessed are the local government and individual employee judgments against Allied. The Environmental Protection Agency reports that "cleaning Kepone out of the James River could cost up to $7 billion" (Chemecology, 1978). We limited our cost estimates, however, to the $2.7 million fisheries losses (Table 5).

A partial ban on fishing was recently imposed in Lake Ontario because of mirex and PCB contamination. This reduced fishing to less than one third of earlier estimates (Brown, 1976) and resulted in an annual loss calculated to be at least $2 million (Table 5).

Birds and mammals in the wild also suffer from exposure to pesticides. Deleterious effects include: death from direct exposure to high dosages; reduced survival, growth and

Table 5. Fishery and wildlife losses due to pesticides.

Fishery Loss	Total Cost
Direct fish kills	$800,000
Lake Ontario fishing restriction	2,000,000
Kepone contamination of James River fishery	2,670,000
Pesticide monitoring of wildlife	5,000,000
Re-establishment of endangered species	250,000
TOTAL	$10,720,000

reproduction from exposure to sublethal dosages; and habitat reduction through elimination of food sources. Because scant data exist on wildlife kills caused by pesticides, it is impossible to estimate mortality rates. There is good reason for this, as terrestrial wildlife are often secretive, camouflaged, highly mobile, and do not conspicuously "float to the surface", as fish do. Even in controlled studies researchers have had great difficulty in finding poisoned birds and mammals (Rosene and Lay, 1963).

Perhaps the various sublethal effects caused by continuous exposure to low-level dosages of pesticides are the most serious problem. Numerous studies have documented that sublethal residues can increase susceptibility to disease, starvation, and other environmental stresses (Friend and Trainer, 1970; Pimentel, 1971; Hill et al., 1971; Scott et al., 1975; Babcock and Flickinger, 1977). Reduced reproductive success in many species of birds, disrupted metabolic processes such as vitamin A utilization, and behavioral changes such as delayed migratory activity have all been linked to low-level pesticide exposure (Stickel, 1973; Jefferies, 1975; Mahoney, 1975).

Although no cost is here assessed for such reductions of any wildlife populations, there is certainly some economic impact to activities concerned with wildlife. But in addition there are aesthetic costs, due to the great value wildlife have for a large segment of the public. Such noneconomic costs certainly cannot be quantitatively assessed in any way here, but it is important that they not be dismissed as insignificant.

Damage to wildlife is also of serious concern because wildlife play an integral role in our ecosystem, which is essential to human survival. Furthermore, wildlife serve as an "early warning system" to alert us of the presence of severe pesticide pollution. Not to be discounted is the important role wildlife play in the nation's economy. The U.S. Bureau of Sport Fisheries and Wildlife documented that fishermen and hunters spend large sums of money pursuing their sport; 36 million fishermen and hunters spent about $7 billion in 1970 (USDI, 1970).

Another cost of pesticide use, which can be directly quantified, is related to the activities of the U.S. Fish and Wildlife Service. They operate a monitoring program specifically concerned with the impact of environmental contaminants on nontarget species. From 60 to 70% of the contaminants monitored are agricultural chemicals and the program has an annual budget of nearly $10 million annually (Shepard, 1978). We assumed that half of this cost could be attributed to pesticides (Table 5).

Despite the difficulty in quantifying wildlife losses there is considerable evidence that pesticides have significantly reduced populations of such bird species as the bald eagle and peregrine falcon (Pimentel, 1971; Stickel, 1973; Edwards, 1973; Brown, 1978a). No attempt has been made to place an economic value on this kind of wildlife loss.

The U.S. Fish and Wildlife Service also spends $250,000 annually in their Endangered Species Program, which aims to re-establish species such as the bald eagle, peregrine falcon, osprey, and brown pelican whose numbers have been severely reduced by pesticides (Shepard, 1978). This too increases the ultimate cost of pesticides.

The fishery and wildlife losses that could be estimated were only $10.7 million (Table 5); most of the costs cannot be calculated.

Invertebrates and Microorganisms

Perhaps more important than the effects on fish and wildlife are the effects of pesticides on insects, earthworms, fungi, bacteria, and protozoa found in soils. These organisms are essential to the proper functioning of all ecosystems since they break down wastes, permitting the vital chemical nutrients to be recycled in the life system. Bacteria and fungi make nitrogen and other elements available to plants. Earthworms and insects aid in turning over the soil at a rate of about 20 tons per acre per year (Kevan, 1962; Burges and Raw, 1967).

These ecological systems are very poorly understood,
but there are specific studies demonstrating the impact of
pesticides on soil organisms. For instance, the pesticides
2,4-D, TCA, CDAA, chlorpropham, monuron, chloronitropropane
and dicloran have all been shown to inhibit nitrification in
soil (Pimentel, 1971); and this effect can be severe enough
to significantly reduce crop yields (Dubey, 1970). However,
no quantitative data exist concerning the overall impact of
pesticides on soil organisms and what this may mean economically to the environment, agriculture, and society as a
whole (Edwards, 1973; Alexander, 1977; Brown, 1978a). Therefore, we can assess no cost to this potentially significant
indirect loss from pesticide use.

Expenses for Government Pesticide Pollution Control

Government expenses constitute a major indirect cost of
pesticide use that is easily overlooked. These are for
state and federal regulatory and monitoring activities
necessitated by the hazards of pesticides. These funds are
spent to reduce the hazards of pesticides, specifically to
protect public health and the integrity of the environment.
They encompass such things as programs each year to train
and license pesticide applicators, and register pesticides.

These pesticide pollution control activities cost the
federal government about $70 million annually (USBC, 1977).
In addition, the individual states have extensive activities
of their own. If the states spend an amount equal to that
spent on the federal level (USBC, 1977), then the governments
together spend about $140 million for pesticide pollution
control.

These relatively high government expenditures for
pesticide regulation and monitoring are aimed at preventing
still higher losses that would be incurred as other types of
indirect costs. In other words, the government is assuming
part of the costs of pesticide use that would otherwise be
suffered by other sectors of the economy. Therefore these
government expenses are an indirect cost of current pesticide
use. They are also an external cost, since they are not
borne by those applying the pesticides.

Insofar as these government programs are effective in
reducing other external costs of pesticides, they may produce
an actual decrease in the total external costs of pesticides.
We are not in a position to judge the appropriateness of any
specific government pollution control program. However, we
do consider it proper that the government spend what it can

on prevention of pesticide damages when these expenses can produce sufficient reductions in losses to other sectors not responsible for the pesticide applications.

Conclusion

Every year in the United States about 1 billion pounds of pesticides are applied, at a direct cost of about $2.8 billion. Of this, about 800 million pounds are applied to crops at a cost of about $2.2 billion. This agricultural usage produces an estimated benefit of $8.7 billion in reduced crop losses (Pimentel et al., 1978a). If this rate of returns on investment is extended to the additional 200 million pounds of pesticide not used on crops, the annual benefits of pesticide use total an estimated $10.9 billion. The $2.8 billion investment that produces this return does include some of the indirect costs considered in this paper -- those for loss of natural enemies, insecticide resistance, and applicator's insurance.

In this preliminary study we estimate the annual indirect costs of pesticide use at about $840 million (Table 6), in those environmental and social effects for which quantitative data are available. This figure includes about $300 million for natural enemy losses, resistance, and insurance costs. However, the remaining $540 million is in external costs not borne by the individuals applying the pesticides. Adding this external cost to the direct cost of $2.8 billion suggests a return of at best about $3 per dollar invested in pesticidal controls, rather than the $4 return calculated by Pimentel et al. (1978a).

The apparent benefit/cost ratio of 4:1 is analogous to the information that farmers must weigh in making their treatment decisions. By the nature of the external costs considered in this paper, they will not be taken into account by those who decide on pesticide use. Thus, cases are to be expected where the best decision will be to apply pesticides even though such an application will be a net harm to society; this will occur because the costs are not borne by those who make the decision to use a pesticide.

We believe that it is economically and morally justified to find some means of transferring these costs to those who use pesticides and reap the benefits. It is recognized that transferring the environmental and social costs of pesticides to the user is difficult, but legislation is needed to implement this transfer.

Table 6. Total estimated environmental and social costs for pesticides in the United States.

Environmental Factor	Total Cost
Human pesticide poisonings	$184,000,000
Animal pesticide poisonings and contaminated livestock products	12,000,000
Reduced natural enemies and pesticide resistance	287,000,000
Honey bee poisonings and reduced pollination	135,000,000
Losses of crops and trees	70,000,000
Fishery and wildlife losses	11,000,000
Government pesticide pollution controls	140,000,000
TOTAL	$839,000,000

Calculating environmental and social costs of pesticides is itself difficult. Our preliminary assessment of the environmental and social costs of pesticides is obviously an oversimplified and incomplete assessment of the existing situation. The immeasurables are replete. For example, it is impossible to place an acceptable monetary value on human lives lost. In addition, many values are unmeasured. A more complete accounting of indirect costs would include additional data about: the total costs of accidental releases of pesticides like the recent Kepone incident; livestock poisonings; pollination losses in crop production; unrecorded losses of fish, wildlife, crops, and trees; losses resulting from the destruction of soil invertebrates, microflora, and microfauna; and chronic health problems like teratogenic and mutagenic effects.

In addition to obtaining more complete data nationally on the costs of pesticides, a need exists for detailed analyses of the costs and benefits of pesticide use regionally throughout the nation. Careful cost/benefit assessments such as those proposed would identify control programs that are cost effective for society as a whole.

What has been accomplished in our investigation is to give a quantitative estimate to some of the indirect costs of using pesticides. While a complete assessment remains a need, it is evident that the indirect costs are probably on the same order of magnitude as the direct costs. The implication for public policy is that the immeasurables are more important than considered in the past.

What sort of policy changes does the realization of the high environmental and social costs of pesticides suggest? The obvious suggestion is to encourage the more effective use of pesticides. It has been proposed that total pesticide use could be reduced 35-50% without lessening the effectiveness of pest control (PSAC, 1965). Monitoring of pest and natural enemy populations to determine when pesticides should be used as in integrated control programs is one way to reduce pesticide use.

Also, actions are needed to reduce the major indirect costs. For example, human pesticide poisonings, increased control costs, and pollinator losses account for most (about 70%) of the calculable indirect costs of pesticide use in the United States (Table 6). Further efforts are needed to reduce these costs.

Clearly, the results of this preliminary assessment underscore the importance of qualitative considerations in determining pesticide strategies. This study also emphasizes the need and direction of more detailed investigations. Pesticides will continue to be a valuable pest control measure, but a more accurate cost/benefit analysis will be helpful as we endeavor to minimize risks and maximize benefits of pest control strategies for society as a whole.

References

Adkisson, P.L. 1971. Objective uses of insecticides in agriculture. pp. 43-51 in Agricultural Chemicals - Harmony or Discord for Food, People, and the Environment. J.E. Swift, ed. Division of Agricultural Sciences, University of California.

Adkisson, P.L. 1972. The integrated control of the insect pests of cotton. Proc. Tall Timbers Conf. Ecol. Anim. Control Habitat Manage. 4:175-188.

Adkisson, P.L. 1973. The principles, strategies and tactics of pest population regulation and control in major crop ecosystems: The cotton system. pp. 274-284 in Studies in Population Management. P.W. Geier, L.R. Clark, D.J. Anderson, and H.A. Nix, eds. Memoirs of the Ecological Society of Australia. Vol. I.

Adkisson, P.L. 1977. Alternatives to the unilateral use of insecticides for insect pest control in certain field crops. pp. 129-144 in Symposium on Ecology and Agricultural Production. L.F. Seatz, ed. University of Tennessee, Knoxville. 247 pp.

AFS. 1975. Monetary values of fish. American Fisheries Society, The Pollution Committee, Southern Division. 18 pp.

Akesson, N.B., W.E. Yates, and P. Christensen. 1978. Aerial dispersion of pesticide chemicals of known emissions, particle size and weather conditions. Manuscript.

Akins, M.B., L.S. Jeffery, J.R. Overton, and T.H. Morgan, Jr. 1976. Soybean response to preemergence herbicides. Proc. S. Weed Sci. Soc. 29:50.

Alexander, M. 1977. Introduction to Soil Microbiology. 2nd Ed. John Wiley and Sons, New York. 467 pp.

AMA. 1977. Profile of Medical Practice, 1977. American Medical Association. S.R. Henderson, ed. Center for Health Services Research and Development.

Anonymous. 1978. Pet population. Pet Food Industry, April, p. 23.

Arnold, W.E. 1979. Personal communication. South Dakota State University.

ASCS. 1976. Unpublished data. Agricultural Stabilization and Conservation Service, Washington, D.C.

Atkins, E.L. 1977. Personal communication. University of California, Riverside.

Auch, D.E. and W.E. Arnold. 1978. Dicamba use and injury on soybean (Glycine max) in South Dakota. Weed Sci. 26:471-475.

Babcock, M. and E. Flickinger. 1977. Dieldrin mortality of lesser snow geese in Missouri. J. Wildl. Manage. 41(1): 100-103.

Barnes, J.M. 1976. Hazards to people. pp. 180-192 in Pesticides and Human Welfare. D.L. Gunn and J.G.R. Stevens, eds. Oxford University Press, Oxford. 278 pp.

Bartlett, B.R. 1968. Outbreaks of two-spotted spider-mites and cotton aphids following pesticide treatment. I. Pest stimulation vs. natural enemy destruction as the cause of outbreaks. J. Econ. Entomol. 61:297-303.

Bauman, T.T. 1978. Personal communication. Purdue University.

Berry, J.H. 1979. Pesticides and energy utilization. in Pesticides: Role in Agriculture, Health, and Environment. T.J. Sheets and D. Pimentel, eds. Humana Press, Clifton, N.J. In press.

Best, W.R. 1963. Drug associated blood dyscrasias. JAMA 185:286-290.

Bidstrup, P.L., J.A. Bonnell, and A.G. Beckett. 1953. Paralysis following poisoning by a new organic phosphorus insecticide (Mipafox). Brit. Med. J. 1:1068-1072.

Blondell, J. 1978. Personal communication. Environmental Protection Agency.

Bogden, J.D., M.A. Quinones and A. El Nakah. 1975. Pesticide exposure among migrant workers in southern New Jersey. Bull. Environ. Contam. Toxicol. 13:513-517.

Bottrell, D.G. and D.R. Rummel. 1978. Response of Heliothis populations to insecticides applied in an area-wide reproduction diapause boll weevil suppression program. J. Econ. Entomol. 71:87-92.

Brazzel, J.R., L.D. Newsom, J.S. Roussel, C. Lincoln, F.J. Williams, and G. Barnes. 1953. Bollworm and tobacco budworm as cotton pests in Louisiana and Arkansas. La. Agr. Exp. Sta. Tech. Bull. 482. 47 pp.

Brown, A.W.A. 1978a. Ecology of Pesticides. John Wiley & Sons, New York. 525 pp.

Brown, C. 1978b. Personal communication. Washington State Department of Agriculture, Pesticides Branch.

Brown, T.L. 1976. The 1973-1975 salmon runs: New York's Salmon River sport fishery, angler activity, and economic impact. New York Sea Grant Institute Publ., Dept. of Natural Resources, Cornell University. 25 pp.

Burges, A. and F. Raw (eds.). 1967. Soil Biology. Academic Press, New York. 532 pp.

Caldwell, S.T. and M.T. Watson. 1975. Hospital survey of acute pesticide poisoning in South Carolina 1971-1973. J. S.C. Med. Assoc. 71:249-252.

Caldwell, S.T., J.E. Keil, and S.H. Sandifer. 1977. Survey of animal pesticide poisoning in South Carolina 1976. J. Vet. Human Toxicol. 19:166-168.

Cann, H.M., D.S. Neyman, and H.L. Verhulst. 1958. Control of accidental poisoning - a progress report. JAMA 168:717-724.

Carlson, L.A. and B. Kolmodin-Hedman. 1972. Hyper-x-lipoproteinaemia in men exposed to chlorinated hydrocarbon pesticides. Acta Med. Scand. 192:29-32.

Carlson, L.A. and B. Kolmodin-Hedman. 1977. Decrease in alphalipoprotein cholesterol in men after cessation of exposure to chlorinated hydrocarbon pesticides. Acta Med. Scand. 201(4):375-376.

Cassarett, L.J., G.C. Fryer, W.L. Yauger, and H.W. Klemmer. 1968. Organochlorine pesticide residues in human tissues - Hawaii. Arch. Environ. Health 17:306.

Cate, J.R., R.L. Ridgway, and P.D. Lingren. 1972. Effects of systemic insecticides applied to cotton on adults of an Ichneumonid parasite, Campoletis perdistinctus. J. Econ. Entomol. 65:484-488.

CCLRS. 1975. California Fruit and Nut Acreage. California Crop and Livestock Reporting Service. USDA, SRS, Cal.

Dep. Food and Agr., Bur. of Agr. Statistics.

Chang, T.H. and W. Kassen. 1968. Comparative effects of tretamine, tepa, apholate, and their structural analogs on human chromosomes in vitro. Chromosoma 24:314-323.

Chang, W.L. 1965. Comparative study of weed control methods in rice. J. Taiwan Agr. Res. 14(1):1-14.

Chemecology. 1978. Kepone antidote discovered: treated workers said improving. March, p. 4. Manufacturing Chemists Association, Washington, D.C.

Clark, L.C., C.M. Shy, B.M. Most, J.W. Florin, and K.M. Portier. 1977. Cancer mortality and agricultural pesticide use in the southeastern United States. 8th Internatl. Sci. Mtg., Internatl. Epidemiol. Assoc. San Juan, Puerto Rico.

Clark, G.M. 1977. Personal communication. United States Department of Agriculture.

CPSC. 1976. National electronic injury surveillance system of emergency rooms. U.S. Consumer Product Safety Commission. Bureau of Epidemiology, Washington, D.C. 1 p. Mimeo.

Croft, B.A. 1978. Potentials for research and implementation of integrated pest management on deciduous tree-fruits. pp. 101-115 in Pest Control Strategies. E.H. Smith and D. Pimentel eds. Academic Press, New York. 334 pp.

Czeizel, A., T. van Bao, I. Szabo, and P. Ruzicska. 1973. Human chromosome aberrations in acute organic phosphorus acid ester. Mutat. Res. 21:187.

Czeizel, A. and J. Kiraly. 1976. Chromosome examination in workers producing Klorinol^R and Bonivol^R. in The Development of a Pesticide as a Complex Scientific Task. L. Banki, ed. Medicina Press, Budapest.

Dacre, J.C. and R.W. Jennings. 1970. Organochlorine insecticides in normal and carcinogenic human lung tissue. Toxic. Appl. Pharmacol. 17:277.

Daniel-Guido, M. 1978. Personal communication. Nutrition Action, Washington, D.C.

Davies, J.E., J.C. Cassady, and A. Raffonelli. 1973. The pesticide problems of the agricultural worker. pp. 223-

231 <u>in</u> Pesticides and the Environment: A Continuing Controversy. W.B. Deichmann, ed. Intercontinental Medical Book Corporation, New York. 223 pp.

Davis, J.H., J.E. Davies, and A.J. Fisk. 1969. Occurrence, diagnosis, and treatment of organophosphate pesticide poisoning in man. Ann. N.Y. Acad. Sci. 160(I):383-392.

Davis, K.L., J.A. Savage, and P.A. Berger. 1978. Possible organophosphate-induced parkinsonism. J. Nerv. Ment. Dis. 166:222-225.

Day, B.E. 1978. The status and future of chemical weed control. pp. 203-213 <u>in</u> Pest Control Strategies. E.H. Smith and D. Pimentel, eds. Academic Press, New York. 334 pp.

Dille, J.R. and P.W. Smith. 1964. Central nervous system effects of chronic exposure to organophosphate insecticides. Aerosp. Med. 35:475-478.

Dubey, D. 1970. Nitrogen deficiency disease of sugar cane probably caused by repeated pesticide application. Phytopathology 60:485-487.

Dubinin, N.P., Y.A. Mtrofanov, and E.S. Manvilova. 1967. Analysis of the thiotepa mutagenic effect on human tissue culture cells. Izv. Akad. Nauk. SSSR (Biol.) 4:477-488.

Duggan, R.E. and M.B. Duggan. 1973. Pesticide residues in food. pp. 334-364 <u>in</u> Environmental Pollution by Pesticides. C.A. Edwards, ed. Plenum, London. 542 pp.

Duke, W.B. 1977. Personal communication. Cornell University.

Durham, W.F., H.R. Wolfe, and G.E. Quinby. 1965. Organophosphorus insecticides and mental alertness. Arch. Environ. Health 10:55-66.

Edwards, C.A. (ed.) 1973. Environmental Pollution by Pesticides. Plenum, London. 542 pp.

Ehler, L.E. 1972. Natural biological control of cabbage looper populations on cotton in the San Joaquin Valley of California. Ph.D. Dissertation, Dept. Entomol. Sci., Univ. California, Berkeley. 114 pp.

Ehler, L.E. 1977. Natural enemies of cabbage looper on cotton in the San Joaquin Valley. Hilgardia 45:73-106.

Ehler, L.E. and J.C. Miller. 1978. Biological control in temporary agroecosystems. Entomophaga 23:207-212.

Ehler, L.E., K.G. Eveleens, and R. van den Bosch. 1973. An evaluation of some natural enemies of cabbage looper on cotton in California. Environ. Entomol. 2:1009-1015.

Ehrlich, P.R., A.H. Ehrlich, and J.P. Holdren. 1977. Ecoscience: Population, Resources and Environment. W.H. Freeman, San Francisco. 1051 pp.

Elliot, B.R., J.M. Lumb, T.G. Reeves, and T.E. Telford. 1975. Yield losses in weed-free wheat and barley due to post-emergence herbicides. Weed Res. 15:107-111.

Elmore, C.J. 1977. Personal communication. University of California, Davis.

EPA. 1972-1976. Fish kills caused by pollution in 1970-1974. Environmental Protection Agency, Washington, D.C.

EPA. 1974. Strategy of the Environmental Protection Agency for controlling the adverse effects of pesticides. Environmental Protection Agency, Office of Pesticide Programs, Office of Water and Hazardous Materials, Washington, D.C. 36 pp.

EPA. 1976. National study of hospital admitted pesticide poisonings. Epidemiologic Studies Program, Human Effects Monitoring Branch, Technical Services Division, Office of Pesticide Programs, Washington, D.C. April.

Epstein, S.S. and M.S. Legator. 1971. The Mutagenicity of Pesticides: Concepts and Evaluation. MIT Press, Cambridge, Mass. 220 pp.

Eveleens, K.G. 1972. Impact of insecticide applications on natural biological control of the beet armyworm in cotton. Ph.D. Dissertation, Dept. Entomol. Sci., Univ. California, Berkeley. 137 pp.

Eveleens, K.G., R. van den Bosch, and L.E. Ehler. 1973. Secondary outbreak induction of beet armyworm by experimental insecticide applications in cotton in California. Environ. Entomol. 2:497-503.

FA. 1975. Citrus Summary 1975. Florida Dept. Agr. Conserv. Serv., Florida Crop Livestock Rept. Serv., Orlando, Fla.

Falcon, L.A., R. van den Bosch, C.A. Ferris, L.K. Stromberg, L.K. Etzel, R.E. Stinner, and T.F. Leigh. 1968. A comparison of season-long pest control programmes in California during 1966. J. Econ. Entomol. 61:633-642.

Falcon, L.A., R. van den Bosch, J. Gallagher, and A. Davidson. 1971. Investigation of the pest status of Lygus hesperus in cotton in central California. J. Econ. Entomol. 64:56-61.

Feldman, R.J. and H.I. Maibach. 1970. Pesticide percutaneous penetration in man, abstracted. J. Invest. Derm. 54:435.

Fisher, J.R. 1977. Guillain-Barré syndrome following organophosphate poisoning. JAMA 238:1950-1951.

Fowler, D.L. and J.N. Mahan. 1975. The pesticide review 1974. U.S. Dep. Agr., Agr. Stab. Conserv. Serv., Washington, D.C. 58 pp.

Fox, R. 1978. Personal communication. Welch Company.

Friend, M. and D.O. Trainer. 1970. Polychlorinated biphenyl: interaction with duck hepatitis virus. Science 170: 1314-1316.

Frisbie, R.E. and J.K. Walker. 1979. Pest management systems for cotton insects. in Pest Management. D. Pimentel, ed. CRC Handbook Series in Agriculture, CRC Press, West Palm Beach, Florida. In press.

FWPCA. 1967-1970. Fish kills by pollution in 1966-1969. Federal Water Pollution Control Administration, Washington, D.C.

Gehlbach, S.H., W.A. Williams, J.S. Woodall, and J.I. Freeman. 1974. Pesticides and human health - an epidemiologic approach. Health Serv. Rep. 89(3):274-277.

Georghiou, G.P. and C.E. Taylor. 1977. Pesticide resistance as an evolutionary phenomenon. pp. 759-785. in Proc. XV Internatl. Congr. Entomol.

Gilley, W. 1978. Personal communication. Virginia Department of Public Health.

Goulding, R. 1969. Pesticide residues as a health hazard - United Kingdom viewpoint. Can. Med. Assoc. J. 100:197-204.

Grignac, P. 1978. The evolution of resistance to herbicides in weedy species. Agro-Ecosystems 4:377-385.

Gumenniyi, V.S. and L.F. Tkach. 1976. [Findings on incidence of diseases of the cardiovascular system and respiratory organs in areas with intense and limited use of pesticides (based on medical examination data).] Gig Sanit. (5):114. Pesticide Abstracts 10:77-0802.

Gutierrez, A.P., L.A. Falcon, W. Loew, P.A. Leipzig, and R. van den Bosch. 1975. An analysis of cotton production in California: a model for Acala cotton and the effects of defoliators on its yields. Environ. Entomol. 4:125-136.

Hahn, R.R. 1977. Personal communication. Cornell University.

Hayes, W.J. 1960. Pesticides in relation to public health. Ann. Rev. Entomol. 5:379-404.

Hayes, W.J. 1964. Occurrences of poisonings by pesticides. Arch. Environ. Health 9:621-625.

Hayes, W.J. 1969. Pesticides and human toxicity. Ann. N.Y. Acad. Sci. 160(I):40-54.

Hayes, W.J., Jr. and W.K. Vaughn. 1977. Mortality from pesticides in the United States in 1973 and 1974. Toxicol. Appl. Pharmacol. 42:235-252.

Henderson, J. 1968. Legal aspects of crop spraying. Univ. Ill. Agr. Exp. Sta. Circ. 99.

HEW. 1960-1966. Pollution-caused fish kills in 1960(-1965). U.S. Dept. of Health, Education and Welfare, Publ. Health Serv. Publ. No. 847.

HEW. 1969. Report of the Secretary's Commission on Pesticides and their Relationship to Environmental Health. U.S. Dep. Health, Education and Welfare, U.S. Govt. Print. Off., Washington, D.C. 677 pp.

Higbee, F. 1978. Personal communication. National Agricultural Aviation Association.

HII. 1976. Sourcebook of Health Insurance Data, 1974-1975. Health Insurance Institute, New York, New York. p. 55.

Hill, E.F., W.E. Dale, and J.W. Miles. 1971. DDT intoxication in birds: subchronic effects and brain residues.

Toxicol. Appl. Pharmacol. 20(4):502-514.

Hoopingarner, R. and A.W. Bloomer. 1970. Lymphocyte chromosome analysis of pesticide exposed individuals. Internatl. Congr. Plant Protection (Paris) 7:772.

Howitt, R.E. 1975. Pesticide externality policy and optimal control approach. Unpublished Ph.D. Thesis, Univ. of California, Davis.

Hurst, H. 1978. Personal communication. Mississippi State University, Delta Branch Experiment Station.

Ignoffo, C.M., D.L. Hostetter, C. Garcia, and R.E. Pinnell. 1975. Sensitivity of the entomopathogenic fungus Nomuraea rileyi to chemical pesticides used on soybeans. Environ. Entomol. 4:765-768.

ILL DEC. 1976. Standard price list of fish for Illinois pollution fish kills. (September 1976). Illinois Dep. Environ. Conserv. Mimeo. 8 pp.

Jefferies, D.J. 1975. The role of the thyroid in the production of sublethal effects by organochlorine insecticides and polychlorinated biphenyls. pp. 131-230 in Organochlorine Insecticides: Persistent Organic Pollutants. F. Moriarty, ed. Academic Press, New York. 302 pp.

Jenkins, R.B. and J.F. Toole. 1964. Polyneuropathy following exposure to insecticides. Arch. Intern. Med. 113:691-695.

Johansen, C.A. 1977. Pesticides and pollinators. Annu. Rev. Entomol. 22:177-192.

Johnson, E.K., J.H. Young, D.R. Molnar, and R.D. Morrison. 1976a. Effects of three insect control schemes on populations of cotton insects and spiders, fruit damage, and yield of Westburn 70 cotton. Environ. Entomol. 5:508-510.

Johnson, D.W., L.P. Kish, and G.E. Allen. 1976b. Field evaluation of selected pesticides on the natural development of the entomopathogen, Nomuraea rileyi, on the velvetbean caterpillar in soybean. Environ. Entomol. 5:964-966.

Keil, J.E., S.H. Sandifer, and R.H. Godsden. 1970. Pesticide poisonings in South Carolina. J. S.C. Med. Assoc. 66: 69-70.

Keil, J.E., J.F. Finklea, S.H. Sandifer, and M.C. Miller. 1972a. Pesticide Exposure Index (PEI). In Determination of Air Quality. G. Mamantov and W.D. Shults, eds. Plenum, New York.

Keil, J., S.H. Sandifer, J.H. Finklea, and L.E. Priester. 1972b. Serum vitamin A elevation in DDT-exposed volunteers. Bull. Environ. Contam. Toxicol. 8:317-320.

Kevan, D.K.McE. 1962. Soil Animals. Philosophical Library, New York. 237 pp.

Kevan, P.G. 1975. Forest application of insecticide fenitrothion and its effect on wild bee pollinators (Hymenoptera: Apoidea) of low-bush blueberries (Vaccinium spp.) in southern New Brunswick, Canada. Biol. Conserv. 7:301-309.

Kim, Y.I., D.W. Whang, Y.I. Nam, C. Lee, H.C. Lee, M.H. Park and H.S. Park. 1977. [Study on liver injury in acute drug intoxification] Taehan Naekwa Hakkoe Chapchi (Kor. J. Intern. Med.) 20(2):146-152 (in Korean).

Kinzer, E.E., C.B. Cowan, R.L. Ridgway, J.W. Davis, J.R. Coppedge, and S.L. Jones. 1977. Populations of arthropod predators and Heliothis spp. after applications of aldicarb and monocrotophos to cotton. Environ. Entomol. 6:13-16.

Kiraly, J., A. Czeizel, and I. Szentesi. 1977. Genetic study on workers producing organophosphate insecticides. Mutat. Res. 46(3):224.

Kiraly, J., I. Szentesi, M. Ruzicska, and A. Czeizel. 1979. Chromosome studies in workers producing organophosphate insecticides. Arch. Environ. Contam. Toxicol. (in press).

Knake, E.L. 1977. Personal communication. University of Illinois.

Komarova, L.I. 1976. [Concerning the influence of chloroorganic pesticides on the development of blood system diseases.] Probl. Gematol. Pereliv. Krovi 21(11): 46-50. Pest. Abst. 10:77-1032.

Koos, B.J. and L.D. Longo. 1976. Mercury toxicity in the pregnant woman, fetus and newborn infant. Am. J. Obstet. Gynecol. 126:390-409.

Kraybill, H.F. 1977. Evaluation of Agricultural Chemicals in the National Cancer Program. Presented at Entomological Society of America, Section E Symposium. November 27-December 1, 1977. Washington, D. C.

Kutz, F.W., S.C. Strassman, and A.R. Yobs. 1977. Survey of pesticide residues and their metabolites in humans. pp. 523-539 in Pesticide Management and Insecticide Resistance. D.L. Watson and A.W.A. Brown, eds. Academic Press, New York. 638 pp.

Lande, S.S. 1974. An epidemiological study of pesticide exposures in Alleghany County, Pennsylvania. Arch. Environ. Health 29:90-95.

Laster, M.L. and J.R. Brazzel. 1968. A comparison of predator populations in cotton under different control programs in Mississippi. J. Econ. Entomol. 61:714-719.

Levin, M.D. 1970. The effects of pesticides on beekeeping in the United States. Am. Bee J. 110:8-9.

Lichtenberg, J.J., J.W. Eichelberger, R.C. Dressman, and J.E. Longbottom. 1970. Pesticides in surface waters of the United States. A five year summary, 1964-1968. Pestic. Monit. J. 4(2):485-488.

Lingren, P.D., R.L. Ridgway, C.B. Cowan, Jr., J.W. Davis, and W.C. Watkins. 1968. Biological control of the bollworm and the tobacco budworm by arthropod predators affected by insecticides. J. Econ. Entomol. 61:1521-1525.

Lingren, P.D., D.A. Wolfenbarger, J.B. Nosky, and M. Diaz, Jr. 1972. Response of Campoletis perdistinctus and Apanteles marginiventris to insecticides. J. Econ. Entomol. 65: 1295-1299.

Lisella, F.S., W. Johnson, and C. Lewis. 1975. Health aspects of organophosphate insecticide usage. J. Environ. Health 38:119-122.

Lopinot, A.C. 1971. Procedures used by the Illinois DEC in the investigation of pollution caused fish kills. Fish Kill Investigation Seminar, in Cincinnati, Ohio, January 12-14, 1971. Collection of Papers. Environmental Protection Agency.

Luckmann, W.H. 1978. Insect control in corn - practices and prospects. pp. 137-155 in Pest Control Strategies. E. H. Smith and D. Pimentel, eds. Academic Press, New York. 334 pp.

Maddy, K.T. 1978. Personal communication. California Dep.
    Food and Agriculture.

Mahadevan, V. and K.C. Chandy. 1959. Preliminary studies
    on the increase in cotton yield due to honeybee pollina-
    tion. Madras Agr. J. 46:23-26.

Mahoney, J.J. 1975. DDT and DDE effects on migratory con-
    dition in white throated sparrows. J. Wildl. Manage.
    39(3):520-527.

Martin, E.C. 1977. Personal communication. United States
    Department of Agriculture.

Martin, E.C. 1978. Impact of pesticides on honeybees.
    Gleanings in Bee Cult. 106:318-320, 346.

Maugh, T.H. 1973. DDT: An unrecognized source of PCB's.
    Science 180:578-579.

Maylin, G. 1977. Personal communication. Cornell University.

McGregor, S.E. 1973. Insect pollination - significance and
    research needs. Am. Bee J. 113:249.

McGregor, S.E. 1976. Insect pollination of cultivated crop
    plants. Agricultural Handbook No. 496. USDA, Agr. Res.
    Serv.

McGregor, S.E. 1977. Personal communication. United States
    Department of Agriculture.

McGregor, S.E., C. Rhyne, S. Worley, Jr., and F.E. Todd.
    1955. The role of honeybees in cotton pollination.
    Agron. J. 47:23-25.

Mengle, D., W. Hale, and R.T. Rappolt. 1966. Hematologic
    abnormalities and pesticides. Calif. Med. 107:251-253.

Metcalf, D.R., and J.H. Holmes. 1969. EEG, psychological,
    and neurological alterations in humans with organophos-
    phorus exposure. Ann. N.Y. Acad. Sci. 160(I):357-365.

Milby, T.H. 1976. in Pesticide residue hazards to farm
    workers. U.S. Dep. Health, Education, and Welfare,
    Public Health Service Center for Disease Control,
    National Institute of Occupational Safety and Health,
    Utah. May. 5 pp.

Milby, T.H. and W.L. Epstein. 1964. Allergic contact sensi-
    tivity to malathion. Arch. Environ. Health 9:434-437.

Nag, D., G.C. Singh, and S. Senon. 1977. Epilepsy epidemic due to benzahexachlorine. Trop. Geog. Med. 29:229-232.

NAS. 1975. Pest Control: An Assessment of Present and Alternative Technologies. Vols. I-V. National Academy of Sciences, Washington, D.C.

Nater, J.P. and V.H.J. Gooskens. 1976. Occupational dermatosis due to a soil fumigant. Contact Derm. 2(4):227-229.

Newsom, L.D. 1962. The boll weevil problem in relation to other cotton insects. pp. 83-94 in Proc. Boll Weevil Research Symp., State College, Miss.

Nora, J.J., A.H. Nora, R.J. Sommerville, R.M. Hill, and D.G. McNamara. 1967. Maternal exposure to potential teratogens. JAMA 202:1065-1069.

NTSB. 1977. Aircraft Accident Reports. Vol. 1-4 for 1976 accidents. National Transportaton Safety Board. U.S. Govt. Print. Off., Washington, D.C.

NYS DEC. 1977-78. Monthly report on toxic substances impacting on fish and wildlife. New York State Dep. Environ. Conserv., Div. Fish Wildl. April-Feb.

Ogawa, J.M., J.D. Gilpatrick and L. Chiarappa. 1976. Review of plant pathogens resistant to fungicides and bactericides. FAO Plant Protection Bulletin, Working Paper presented at FAO Panel of Experts on Pest Resistance to Pesticides and Crop Loss Assessment, Aug. 1976, Washington, D.C.

Oka, I.N. and D. Pimentel. 1974. Corn susceptibility to corn leaf aphids and common corn smut after herbicide treatment. Environ. Entomol. 3(6):911-915.

O'Leary, J.A., J.E. Davies, W.F. Edmundson, and G.A. Reich. 1970. Transplacental passage of pesticides. Am. J. Obstet. Gynecol. 107:65-68.

OSHA. 1978. Statement of Samuel Epstein, M.D. OSHA Docket No. 090, Occupational Safety and Health Administration, Washington, D.C.

Owens, C.B., E.W. Owens, and D. Zahn. 1978. The extent of exposure of migrant workers to pesticide and pesticide residues. Internatl. J. Chronobiol. 5(2):428-429.

Pate, T.L., J.J. Hefner, and C.W. Neeb. 1972. A management program to reduce cost of cotton insect control in the Pecos area. Texas Agr. Exp. Sta. Misc. Publ. 1023.

Peters, D.C., E.A. Wood, Jr., and K.J. Starks. 1975. Insecticide resistance in selections of the greenbug. J. Econ. Entomol. 68:339-340.

Pilinskaya, M.A. 1970. Chromosome aberrations in the persons contacted with Ziram. (in Russian, English summary). Genetika 6(7):157-163.

Pimentel, D. 1971. Ecological Effects of Pesticides on Non-target Species. U.S. Govt. Print. Off., Washington, D.C. 220 pp.

Pimentel, D. 1977. Ecological basis of insect pest, pathogen and weed problems. pp. 3-31 <u>in</u> The Origins of Pest, Parasite, Disease and Weed Problems. J.M. Cherrett and G.R. Sagar, eds. Blackwell Scientific Publ., Oxford. 413 pp.

Pimentel, D., E.C. Terhune, R. Dyson-Hudson, S. Rochereau, R. Samis, E. Smith, D. Denman, D. Reifschneider, and M. Shepard. 1976. Land degradation: effects on food and energy resources. Science 194:149-155.

Pimentel, D., C. Shoemaker, E.L. LaDue, R.B. Rovinsky, and N.P. Russell. 1977a. Alternatives for reducing insecticides on cotton and corn: economic and environmental impact. Report on Grant No. R802518-02, Office of Research and Development, Environmental Protection Agency.

Pimentel, D., E. Terhune, W. Dritschilo, D. Gallahan, N. Kinner, D. Nafus, R. Peterson, N. Zareh, J. Misiti, and O. Haber-Schaim. 1977b. Pesticides, insects in foods, and cosmetic standards. BioScience 27:178-185.

Pimentel, D., J. Krummel, D. Gallahan, J. Hough, A. Merrill, I. Schreiner, P. Vittum, F. Koziol, E. Back, D. Yen, and S. Fiance. 1978a. Benefits and costs of pesticide use. BioScience 28:772, 778-784.

Pimentel, D., C. Shoemaker, R.J. Whitman, A. Bellotti, N. Beyer, A. Brick, C. Brodel, I. Caunter, H. Cornell, W. Dritschilo, D. Gunnison, M. Habte, L. Hurd, P. Johnson, J. Krummel, J. Liebherr, M. Loye, D. Mackenzie, D. Nafus, I. Oka, R. Rao, D. Saari, J. Smith, R. Stack, D. Udovic, C. Yip and N. Zareh. 1978b. Systems management program for corn pest control in New York State. Search, Cornell University Res. Publ. 8(1):1-16.

Plapp, F.W. and S.B. Vinson. 1977. Comparative toxicities of some insecticides to the tobacco budworm and its ichneumonid parasite, Campoletis sonorensis. Environ. Entomol. 6:381-384.

Plowright, R.C., B.A. Pendrel and I.A. McLaren. 1978. The impact of aerial fenitrothion spraying upon the population biology of bumble bees (Bombus Latr.: Hym.) in southwestern New Brunswick. Can. Entomol. 110:1145-1156.

Potashnik, G., N. Ben-Aderet, R. Israeli, I. Yanai-Inbur, and I. Sober. 1978. Suppressive effect of 1,2-Dibromo-3-chloropropane on human spermatogenesis. Fert. Ster. 30: 444-447.

PSAC. 1965. Restoring the Quality of our Environment. Rep. Environmental Pollution Panel, Pres. Sci. Adv. Comm., The White House.

Quinones, M.A., J.D. Bogden, D.B. Louria, A.E. Nakah, and C. Hansen. 1976. Depressed cholinesterase activity among farm workers in New Jersey. Sci. Total Environ. 6(2): 155-159.

Radomski, J.L., W.B. Deichmann, E.E. Clizer, and A. Rey. 1968. Pesticide concentrations in the liver, brain, and adipose tissue of terminal hospital patients. Food Cosmet. Toxicol. 6:209-220.

Ramsay, R.C., A.S. Fitzhugh, P.C. White, and H.R. Ellis. 1976. Arkansas animal morbidity report. 44 pp.

Reich, G.A., J.H. Davis, and J.E. Davies. 1968a. Pesticide poisoning in South Florida: an analysis of mortality and morbidity and a comparison of sources of incidence data. Arch. Environ. Health 17:768-775.

Reich, G.A., G.L. Gallagher, and J.S. Wiseman. 1968b. Characteristics of pesticide poisoning in South Texas. Tex. Med. 64:56-58.

Reynolds, H.T. 1977. Problems of resistance in pests of field crops. pp. 794-799 in Proc. XV. Internatl. Congr. Ent.

Rhoads, S.E. 1978. How much should we spend to save a life? Public Interests 51:74-92.

Richardson, J.W. 1973. Environ-economic analysis of present and alternative methods of pest management on selected

Oklahoma crops. Unpublished M.S. thesis, Oklahoma State University.

Ridgway, R.L. and P.D. Lingren. 1972. Predaceous and parasitic arthropods as regulators of Heliothis populations. Southern Coop. Serv. Bull. No. 169:48-56.

Ridgway, R.L., P.D. Lingren, C.B. Cowan, and J.W. Davis. 1967. Populations of arthropod predators and Heliothis spp. after applications of systemic insecticides to cotton. J. Econ. Entomol. 60:1012-1096.

Rosene, W. and D.W. Lay. 1963. Disappearance and visibility of quail remains. J. Wildl. Manage. 27(1):139-142.

Sandifer, S.H. and J.E. Keil. 1971. Pesticide exposure: association with cardiovascular risk factors. pp. 329-333 in Trace Substances in Environmental Health - V. A Symposium. D.D. Hemphill, ed. University of Missouri, Columbia, MO.

Schiermeyer, G. 1977. Personal communication. United States Department of Agriculture.

Schotterfeld, D. 1978. Personal communication. Sloan-Kettering Cancer Center.

Schulze, J.A., D.B. Manigold, and F.L. Andrews. 1973. Pesticides in selected western streams - 1968-1971. Pestic. Monit. J. 7(1):73-84.

Scott, J.M., J.A. Wiens, and R.R. Claeys. 1975. Organochlorine levels associated with a common murre die-off in Oregon. J. Wildl. Manage. 39(3):310-320.

Scott, R. 1978. Reproductive hazards. Job Safety Health 6(5):7-13.

Shepard, J.M. 1978. Personal communication. U.S. Fish and Wildlife Service.

Sherry, D. 1971. Fish kill damages; monetary values of fishes. Fish Kill Investigation Seminar, January 12-14, 1971. Collection of Papers. Environmental Protection Agency, Cincinnati, Ohio.

Shishikin, E.A. 1946. [Honeybees in the service of cotton pollination.] Pchelovodstvo 23(5-6):31-32 [In Russian.] Biol. Abstr. 21:26151. p. 2551, 1947.

Slife, F.W. 1972. Personal communication. University of Illinois.

Smith, D.A. and J.S. Wiseman. 1971. Pesticide poisoning epidemiology of pesticide poisoning in the lower Rio Grande Valley in 1969. Tex. Med. 67(2):56-59.

Stanley, C.W., J.E. Barney, M.R. Helton, and A.R. Yobs. 1971. Measurement of atmospheric levels of pesticides. Environ. Sci. Tech. 5:430-435.

Starr, H.G. and N.J. Clifford. 1971. Absorption of pesticides in a chronic skin disease. Arch. Environ. Health 22:396-400.

State of California. 1974. Occupational disease in California attributed to pesticides and other agricultural chemicals, 1970-1973. Dept. of Health, Occupational Health Section and Center for Health Statistics, Berkeley, CA.

Stern, V.M. 1976. Ecological studies of lygus bugs in developing a pest management program for cotton pests in the San Joaquin Valley, California. pp. 8-11 in Lygus Bug: Host Plant Interactions. D.R. Scott and L.E. O'Keefe, eds. Proc. Workshop XV Internatl. Cong. Entomol.

Stickel, L.F. 1973. Pesticide residues in birds and mammals. pp. 254-312 in Environmental Pollution by Pesticides. C.A. Edwards, ed. Plenum, London. 542 pp.

Stoller, A., J. Krupinski, A.J. Christophers, and G.K. Blanks. 1965. Organophosphorus insecticides and major mental illness. Lancet 1:1387-1388.

Stricker, M.H. 1977. Blueberry pollination, 1977. Gleanings in Bee Cult. 105:49-50.

Swartz, J. 1974. Poisoning farmworkers. Environment 17(4): 26-33.

Sweet, R.D. 1977. Personal communication. Cornell University.

Tabershaw, I.R., and W.C. Cooper. 1966. Sequelae of acute organic phosphate poisoning. J. Occup. Med. 8:5-20.

Takahasi, W., A.M. Budy, and L. Wong. 1978. Occurrence of leukopenia among workers exposed to pesticides. Pharmacologist 20:212.

Taylor, R.C. and J.C. Headley. 1975. Insecticide resistance and the evaluation of control strategies for an insect population. Can. Entomol. 107:237-242.

Teetes, G.L., G.A. Schaefer, J.R. Gipson, R.C. McIntyre, and E.E. Latham. 1975. Greenbug resistance to organophosphorus insecticides on the Texas High Plains. J. Econ. Entomol. 68:214-216.

Turner, S. 1978. Personal communication. Stuart Turner and Company.

USBC. 1973a. Census of Agriculture, 1969. Vol. 5. Special Reports. Part 1. Grains, Soybeans, Dry Beans, Dry Peas. U.S. Bureau of the Census. U.S. Govt. Print. Off., Washington, D.C.

USBC. 1973b. Census of Agriculture, 1969. Vol. 5. Special Reports. Part 4. Sugar Crops, Potatoes, Other Specified Crops. U.S. Govt. Print. Off., Washington, D.C.

USBC. 1973c. Census of Agriculture, 1969. Vol. 5. Special Reports. Part 5. Vegetables, Including Tomatoes and Melons. U.S. Govt. Print. Off., Washington, D.C.

USBC. 1977. Statistical Abstract of the United States 1977. U.S. Bureau of the Census, 98th ed. U.S. Govt. Print. Off., Washington, D.C. 1048 pp.

USDA. 1971. The pesticide review 1970. Agr. Stab. Conserv. Serv., Washington, D.C. 46 pp.

USDA. 1975a. Farmers' use of pesticides in 1971 ... extent of crop use. Econ. Res. Serv., Agr. Econ. Rep. No. 268.

USDA. 1975b. Agricultural Statistics 1975. U.S. Govt. Print. Off., Washington, D.C. 621 pp.

USDA. 1975c. Nutritive Value of American Foods in Common Units. Agr. Res. Serv., Agr. Handbook No. 456. 291 pp.

USDA. 1975d. Farmers' use of pesticides in 1971 ... expenditures. Econ. Res. Serv., Agr. Econ. Rep. No. 296. 42 pp.

USDA. 1976a. The Pesticide Review 1975. U.S. Dept. Agr., Agr. Stab. Conserv. Serv. 40 pp.

USDA.   1976b.   Agricultural Statistics 1976.   U.S. Govt.
    Print. Off., Washington, D.C.

USDA.   1977.   Agricultural Statistics 1977.   U.S. Govt. Print.
    Off., Washington, D.C.

USDI.   1970.   National survey of fishing and hunting.   U.S.
    Dept. of the Interior.   Res. Publ. 95, Fish and Wildl.
    Serv., Bur. Sport Fish and Wildl.

USDL.   1975.   Handbook of Labor Statistics, 1975.   U.S. Dept.
    of Labor.   Bureau of Labor Statistics.   p. 188.

U.S. Senate.   1976.   Hearings before the Subcommittee on
    Agricultural Research and General Legislation of the
    Committee on Agriculture and Forestry.   United States
    Senate, Ninety-fourth Congress, Second Session, on the
    Kepone Contamination in Hopewell, Virginia.   U.S. Govt.
    Print. Off., Washington, D.C.   427 pp.

van Bao, T., I. Szabo, P. Ruzicska, and A. Czeizel.   1974.
    Chromosome aberrations in patients suffering acute
    organic phosphate insecticide intoxication.   Humangenetik
    24:33-57.

Van Steenwyk, R.A., N.C. Toscano, G.R. Ballmer, K. Kido, and
    H.T. Reynolds.   1975.   Increases of Heliothis spp. in
    cotton under various insecticide treatment regimes.
    Environ. Entomol. 4:993-996.

von Rumker, R. and F. Horay.   1974.   Farmers' pesticide use
    decisions and attitudes on alternate crop protection
    methods.   U.S. Environmental Protection Agency, Washington,
    D.C.

Walker, R.H.   1977.   Personal communication.   Auburn University.

Ward, C.R., E.W. Huddleston, D. Ashdown, J.C. Owens, and
    K.L. Polk.   1970.   Greenbug control on grain sorghum and
    the effects of tested insecticides on other insects.   J.
    Econ. Entomol. 63:1929-1934.

Wassermann, M., D.P. Nogueira, L. Tomatis, A.P. Mirra, H.
    Shibata, G. Arie, S. Cucos, and D. Wassermann.   1976.
    Organochlorine compounds in neoplastic and adjacent
    apparently normal breast tissue.   Bull. Environ. Contam.
    Toxicol. 15:478-483.

Wassermann, M., D.P. Nogueira, S. Cucos, A.P. Mirra, H.
    Shibata, G. Arie, H. Miller, and D. Wassermann.   1978.

Organochlorine compounds in neoplastic and adjacent apparently normal gastric mucosa. Bull. Environ. Contam. Toxicol. 20:544-553.

West, I. 1968. Sequelae of poisoning from phosphate ester pesticides. Ind. Med. Surg. 37:538.

West, I. and T.H. Milby. 1965. Public health problems arising from the use of pesticides. Residue Reviews 11:1965.

Wheater, R.H. 1978. Short-term exposures to pesticide DBCP and male sterility. JAMA 239:2795.

Whitlock, N., J.E. Keil, and S.H. Sandifer. 1972. Pesticide morbidity in South Carolina. J.S.C. Med. Assoc. 68: 109-112.

Whorton, D., R.M. Krauss, S. Marshall, and T. Milby. 1977. Unfertility in male pesticide workers. Lancet 2(8051): 1259-1261.

Wicker, G.W. 1976. in Pesticide residue hazards to farm workers. U.S. Dept. Health, Education and Welfare. Public Health Service Center for Disease Control, National Institute of Occupational Safety and Health, Utah. May. 115 pp.

Wille, J.E. 1951. Biological control of certain cotton insects and the application of new organic insecticides in Peru. J. Econ. Entomol. 44:13-18.

Williams, I.N., H.S. Pepin, and M.J. Brown. 1976. Degradation of carbofuran by soil microorganisms. Bull. Environ. Contam. Toxicol. 15:244-250.

Wittwer, S.H. 1975. Food production: technology and the resource base. Science 188:579-584.

Wolfe, H. 1976. in Pesticide residue hazards to farm workers. U.S. Dept. Health, Education and Welfare, Public Health Service Center for Disease Control, National Institute of Occupational Safety and Health, Utah. May. 13 pp.

Wolfe, N.L., R.G. Zepp, J.A. Gordon, and R.C. Fincher. 1976. N-Nitrosamine formation from atrazine. Bull. Environ. Contam. Toxicol. 15:342-347.

Yates, W.E. and N.B. Akesson. 1973. Reducing pesticide chemical drift. pp. 275-341 <u>in</u> Pesticide Formulations. Marcel Dekker, New York.

Yoder, J., M. Watson, and W.W. Benson. 1973. Lymphocyte chromosome analysis of agricultural workers during extensive occupational exposure to pesticides. Mutat. Res. 21:335-340.

Young, W.R. and G.L. Teetes. 1977. Sorghum entomology. Ann. Rev. Entomol. 22:193-218.

John Krummel, Judith Hough

5. Pesticides and Controversies: Benefits versus Costs

Abstract

This paper summarizes an overall analysis of the benefits of pesticide use in the United States. Risk/benefit analysis is considered in general terms, and contrasted with classic cost/benefit methodology. Finally, the kind of information that should be included in a benefit analysis is discussed in relation to a specific pesticide, chlorobenzilate.

Introduction

The number of pesticidal active ingredients deliberately introduced into the environment in the United States now exceeds 1,200 (FCH, 1978). The possible detrimental effects of certain chemical pesticides on human health concerns many researchers (Pimentel et al., 1979a). In addition, the action of pesticides and other chemicals on the nontarget biota could have long-term deleterious effects on ecosystem functions (Woodwell, 1978). Thus, a careful assessment of benefits and risks should be a prerequisite to the continued use of chemical pesticides. We present here an analysis of the benefits of pesticide use, followed by a case study of a benefit analysis applied to a single pesticide, chlorobenzilate.

Benefits of Pesticide Use

There is no doubt that considerable direct dollar benefits are derived from the use of pesticides. Previous analyses have estimated dollar returns at from $3 to $5 for every $1 invested in pesticidal control (PSAC, 1965; Headley, 1968; Pimentel, 1973). Nevertheless, the benefits of pesticides in the U.S. agricultural system are sometimes overstated. For example, a recent USDA publication states that "pesticides have been responsible for much of the yield

gains in modern farm production" (USDA, 1978). Concerning
losses without pesticide use, Norman Borlaug suggested that
if pesticides were completely banned, 50% of current crop
production would be lost, and food prices would increase 4-
to 5-fold (Borlaug, 1972). Statements like this are found
in the popular press as well. In a recent issue of Newsweek
magazine, J. W. Hanley, president and chairman of the board
of Monsanto, quotes U.S. Department of Agriculture sources
as stating that crop production would decline 30% and food
prices go up 75% "if farmers quit using modern pesticides"
(Hanley, 1979).

These estimates are probably serious overstatements,
for several reasons. First, a relatively small percentage
of crop acreage is treated with pesticides; second,
nonchemical pest control practices are currently used
effectively on more acreage than chemical control practices;
and finally, losses to pests are already substantial, even
with current chemical and nonchemical control methods.
Evidence for these statements follows.

Current Use of Pesticides

Since the introduction of the chlorinated hydrocarbons
in 1945, pesticide production has increased dramatically,
and there has been no apparent slowdown in the rate of
increase (USDA, 1978). Presently, over 1 billion pounds of
pesticides are used in the United States, with about 660
million pounds applied to agricultural land (USDA, 1978).
Despite the use of large quantities of pesticides, the
actual percentage of crop acres treated remains small. Only
about 9% of U.S. crop acreage is treated with insecticides,
22% with herbicides, and 1% with fungicides (USDA, 1978).
If agricultural land devoted solely to pastures is discounted,
these figures increase to about 18% of crop acreage treated
with insecticides, 56% with herbicides, and 2% with fungicides. As mentioned by Headley in this volume (Chapter 3),
only about half of all U.S. farmers use any pesticides at
all on their land.

Certain large-acreage crops, such as corn, soybeans,
rice, peanuts, and cotton, have more than 80% of their
acreage treated with herbicides (USDA, 1978). Of the major
food crops grown in the United States, however, only corn
and peanuts have more than 30% of their acres treated with
insecticides. The nonfood crops, tobacco and cotton, have
76% and 60% of their acres treated with insecticides, respectively. Peanuts, tobacco, and certain fruits and vegetables
are the only crops that have over 10% of their acreage
treated with fungicides (USDA, 1978).

The amount of pesticide applied to U.S. crop acres increased 38% from 1971 to 1976 (USDA, 1978). The intensified
use of herbicidal weed control accounted for most of this
increase. Agricultural use of herbicides climbed from 207
million pounds in 1971 to 374 million pounds in 1976 (USDA,
1978). A substantial increase in the acres treated and the
amount of herbicide applied per acre on the nation's corn
crop contributed 64% of this increase. In fact, 57% of the
additional 186 million pounds of all pesticides applied in
1976 as compared to 1971 can be traced to increased herbicide
use on corn. In contrast to herbicide use, the amounts of
insecticide and fungicide applied to crops increased by only
4 and 1.7 million pounds, respectively, over this same time
period (USDA, 1978). Cotton and tobacco accounted for more
than 40% of all insecticides used on farms in 1976 (USDA,
1978). Peanuts, sugar beets, potatoes, and certain fruits
and vegetables used over 95% of all fungicides applied to
crop land.

The percent acreage treated with pesticides for an
individual crop often differs in different growing regions
in the United States. For example, 78% of wheat grown in
the Lake states received herbicide treatments, while only
10% of the wheat acreage in the Southern Plains was treated
(USDA, 1978). Insecticide treatments were applied to 99% of
the cotton acreage in the Delta states, while only 30% of
cotton acreage in the Southern Plains was treated. Also,
48% of the soybean land in the Southeast received 1 or more
insecticide treatments per year, compared to only 1% in the
Corn Belt. In the Southeast, nearly all early potato plan-
tings received at least 1 insecticide treatment, while only
65% of the extensive potato acreage in the Mountain states
was treated (USDA, 1978). Much of this geographical varia-
tion undoubtedly reflects the more favorable pest conditions
that develop in warmer, wetter climates.

Nonchemical Pest Control

The figures cited above refer to chemical control. To
put them in perspective, nonchemical pest controls are
actually used more extensively than chemicals. For insects,
nonchemical controls are widely used on certain large-
acreage crops. For example, corn rootworms are controlled
on about 60% of all corn acreage by crop rotation (Pimentel
et al., 1977a). In addition, over one-third of U.S. corn
acreage, or 21.5 million acres, is planted to varieties that
are resistant to the European corn borer. About 10 million
acres of corn are planted to varieties resistant to the
chinch bug (Schalk and Ratcliffe, 1976). Plant resistance
is also important in the control of insect pests of alfalfa,

barley, and grain sorghum. The major insect pest of wheat,
the Hessian fly, is almost entirely controlled by the use of
resistant varieties and the manipulation of planting date
(PSAC, 1965). Natural enemies are important in the control
of insect pests of many orchard crops, such as citrus and
olives, which are grown on about 2 million acres in the
United States (Sweetman, 1958; van den Bosch and Messenger,
1973). Overall, it is estimated that nonchemical insect
pest control methods are used on about 9% of U.S. crop
acreage (Pimentel, 1976), the same percentage on which
insecticides are used.

Weeds are still controlled on most U.S. crop acreage by
tillage and cultural practices, sometimes in combination
with the use of herbicides (NAS, 1968a). Thus, nonchemical
weed control methods are used on an estimated 80% of all
crop acreages (Pimentel, 1976), while herbicides are used on
only 22% of crop acreage.

For diseases, the primary means of control are nonchemical, especially the use of resistant varieties and cultural
manipulations. Disease-resistant varieties are used on
about 75% of all crop acreage, and most of the major crop
varieties now in use incorporate some degree of resistance
to one or more important diseases (NAS, 1968b). Another
important nonchemical disease control technique is the use
of disease-free propagated material. Thus, most bean, pea,
and potato seed planted in the United States is relatively
disease-free (Pimentel et al., 1979b). Crop rotations are
another very important means of controlling many diseases.
Overall, nonchemical methods of disease control are used on
an estimated 90% of all U.S. crop acreage (Pimentel, 1976),
compared with 2% for fungicides.

Current Crop Losses

An analysis of the benefits of pesticide use must take
into account the fact that large acreages of crops are grown
successfully without pesticides, and that certain nonchemical
methods of pest control are widely and successfully used.
Such an analysis must also take into account the fact that
current crop losses to pests are quite substantial, even
with the use of pesticides and other control methods.
Although it is difficult to estimate losses of potential
crop production, the U.S. Department of Agriculture has
suggested that nationwide about a third of potential production is lost to pests: 13% to insects, 12% to plant pathogens, and 8% to weeds (USDA, 1965; Pimentel, 1976). USDA
survey data from the 1940s to the present suggest that
production losses from weeds have declined over that period,

probably due to improved herbicidal and mechanical control technologies. Losses from plant pathogens have increased slightly. Losses from insects, however, have increased substantially, from about 7% in the 1940s to about 13% today (Table 1).

A number of factors have undoubtedly contributed to this increased loss. A very important factor concerns the substantial changes that have occurred in farming practices during the last 30 years, including large increases in the size of farms, and a considerable decline in labor input (Headley, this volume, Chapter 3). Thus, many crops are now grown in extensive monocultures, and may be more likely to be discovered and heavily damaged by certain insect pests (Pimentel, 1977). Crops are also being grown in new areas, where pest pressure may be greater. For example, since 1961 soybean acreage in the United States has more than doubled, to over 55 million acres, and much of the expansion has occurred in southern states. While insect pest problems are of little importance in the Midwest and North Central states, in the South a large number of pest species attack the crop (Newsom, 1978). Crop breeding is another factor that may have increased losses to pests. Until recently, crop breeding has emphasized yield, so that in some cases varieties have been developed that are more susceptible to insects, while natural resistance has been lost or reduced (Lupton, 1977). In other cases, sanitation, including destruction of crop residues, has been decreased, which can allow greater buildup of insect pest populations. Finally, "cosmetic" standards that emphasize the external appearance of foods, especially fruits and vegetables, have become more stringent in the last 30 years (Pimentel et al., 1977b). For this reason, dollar losses due to insect pests may be greater today, even where actual yield losses have not changed.

An Analysis of Pesticide Benefits

A general analysis of the benefits of chemical pesticides, including current patterns of pesticide use and estimated additional crop losses that would occur if pesticides were no longer used, was recently carried out by an interdisciplinary group of workers at Cornell University (Pimentel et al., 1978, 1979b). For each crop, the following information was sought: acreage grown in the United States; dollar value of the crop; food energy, in kilocalories, of the crop; percent of acreage currently treated with pesticides, and the cost of that treatment; current estimated losses to pests; additional losses that would be incurred if pesticides were no longer used, but if certain readily available alternatives were used; and the cost of using

Table 1. Comparison of annual pest losses (dollars) in the USA for the periods 1904, 1910–1935, 1942–1951, 1951–1960, 1974, plus an estimate of losses if no pesticides were used and some nonchemical alternatives were employed.

Period	Source	Percentage of pest losses in crops				Crop value	
		Insects	Diseases	Weeds	Total	$ \times 10^9$	Source
Without pesticides*	Pimentel et al., 1979b	18.0	15.0	9.0	42.0	77	USDA, 1975a
1974	Pimentel, 1976	13.0	12.0	8.0	33.0	77	USDA, 1975a
1951–1960	USDA, 1965	12.9	12.2	8.5	33.6.	30	USDA, 1961
1942–1951	USDA, 1954	7.1	10.5	13.8	31.4	27	USDA, 1954
1910–1935	Hyslop, 1938	10.5	NA†	NA	NA	6	USDA, 1936
1904	Marlatt, 1904	9.8	NA	NA	NA	4	Marlatt, 1904

* Includes the substitution of some nonchemical alternative controls.

† Not available.

those alternatives. The results of that study, summarized in Table 1, indicate that without insecticides dollar losses would increase by about 5% above current losses to insects. Without herbicides, there would be only a 1% increase in crop losses due to weeds. This is because weed control can be achieved relatively effectively by mechanical and cultural methods, especially in the large-acreage row crops. For diseases, additional crop losses without fungicide use were estimated at about 3%.

Overall, then, the study concluded that dollar crop losses would amount to an estimated total loss of about 9%. Thus, current pest losses (about 33%) would increase to about 42% of potential crop production. If nonfood crops like cotton, tobacco, hay, and pasture, are excluded, the loss estimate increases to 11% of current production. This is considerably lower than the 50% loss forecast by Borlaug (1972), or even the 30% loss quoted by Hanley (1979).

These figures are based on dollar value. As a rough estimate of loss of food energy, we converted our estimates to kilocalories. Total loss on this basis would amount to only about 1% of all crops, or 4% of food crops (Pimentel et al., 1978). These losses are lower than those based on dollar value, because high-calorie crops such as wheat and corn would be less affected by pesticide loss than lower calorie crops such as fruits and vegetables. The contrast between these two estimates points out the difficulties in trying to summarize data for very different kinds of crops, such as apples and field corn. An estimate based strictly on kilocalories probably undervalues the importance of fruits and vegetables in our diet, as they are one of our major sources of essential vitamins and minerals. At the same time, estimates based strictly on dollar value probably overestimate their importance.

The results of this analysis indicate that there would be no serious food shortage in the United States without pesticide use, even with only limited use of available alternative control techniques. The estimates do suggest that serious shortages of certain fruits and vegetables, including apples, peaches, onions, and tomatoes, might occur if pesticides were no longer used. However, in making these estimates, we accepted current grading standards, which in many cases are based at least in part on external appearance of fruits and vegetables. Thus, we most likely would experience a shortage of "perfect" fruits and vegetables rather than a loss of all produce, if pesticide use were restricted. In support of this contention, in 1909, when pesticide use was much less intense than it is today, per capita consumption

of fresh and processed fruits was 130 lb per year, compared
with 136 lb in 1975; consumption of fresh and processed
vegetables was 204 lb, compared with 206 lb in 1975 (USDA,
1966; 1975b). Some of the "loss" predicted by the analysis,
then, can probably be attributed to the more stringent
quality and cosmetic standards that are followed today
(Pimentel et al., 1977b).

Clearly, the U.S. population has increased substantially
since 1909. Although pesticides have undoubtedly helped
food production to keep pace with population growth, other
agronomic practices have also contributed. For example,
Jugenheimer (1976) states that improved hybrids, increased
nitrogen use, and heavy plant populations were the most
important factors in the 2.5 bushels/acre annual increase in
corn yields seen over the last 20 years. He also estimates
that hybrid seed alone increased corn yields in the United
States 25-50%. The increased use of nitrogen over the last
20 years to present levels of 120 lb/acre was also a significant factor in the yield gains (Durost, 1970).

Risk/Benefit Analysis

While a general analysis of the benefits and costs of
pesticide use in the United States is important in assessing
the value of pesticides to agriculture as a whole, decision
makers must also make risk/benefit analyses of specific pesticides, based on detailed scientific data. For example, by
law the Environmental Protection Agency must conduct a risk/
benefit analysis for the reregistration of all pesticides
placed on the Rebuttable Presumption Against Registration
(RPAR) list. These are chemicals that have some risk associated with their use. If the risks are judged to exceed the
benefits, these chemicals will be restricted or banned from
further use.

The use of a risk/benefit analysis in the decision-
making process allows greater flexibility in the interpretation of available data than does a classic cost/benefit
analysis. In the standard cost/benefit analysis, all the
data must be reduced to the same units, most often dollars.
While obvious benefits of a chemical can be measured in the
marketplace, many indirect benefits and costs cannot be
determined in dollars. In some cases, a survey of willing-
ness-to-pay can provide a method for analyzing costs and
benefits. For example, one researcher (Lave, 1972) has
asked, rhetorically, what we would pay to add one year's
life expectancy through a 50% reduction in air pollution
levels. However, an individual's response to such a survey
would be biased by narrow opinions and lack of knowledge

concerning all the benefits and risks (Tihansky and Kibby, 1974). As another example of problems in assigning costs, it is almost impossible to place a dollar value on wildlife, such as birds of prey whose reproductive rate is threatened by a persistent pesticide like DDT. As Tihansky and Kibby (1974) point out, many costs, such as seagull poisonings, can be quantified, but can not be translated into monetary terms.

Ideally, all the risks and benefits that result from the use of pesticides should be identified. However, to achieve this task would be too expensive and time consuming, and often beyond present scientific expertise. With over 1,200 pesticidal ingredients and several thousand more formulations currently on the market, risk/benefit analyses of these chemicals represent a large undertaking. Also, most of these pesticides are used on more than one crop and against more than one pest species, which compounds the difficulty of a risk/benefit analysis. As a pragmatic first step, decision makers must start by dealing with those pesticides judged to have the greatest risk associated with their use.

The problem of risk uncertainty confronts all who attempt a risk/benefit analysis. Problems of biological accumulation, genetic effects, and low-level, long-term dosages create pitfalls in the assessment of the risks of pesticide use (Tihansky and Kibby, 1974). Thus, DDT was used for many years before adequate information was collected to confirm its deleterious long-term ecological effects. Indeed, the assumption can be made that any persistant and/or reactive chemical released into the environment will perturb some ecosystem. The idea of a threshold level of a toxic substance may be a meaningless biological notion (Woodwell, 1978).

The quality of a risk/benefit analysis depends ultimately on the data base. Risk assessments are usually triggered by scientific information indicating detrimental effects on human health or the environment. Economic benefits of a pesticide are usually measured in terms of the chemical's effect on crop production. The economic data are then translated into costs that would be incurred by the farmer, processor, and consumer if restrictions were placed on the use of the pesticide. If the data are based solely on the efficacy of a pesticide, rather than on information about the relationship between crop yield and pest infestation, discrepancies can occur among different estimates of the extent of the benefits (Pimentel et al., 1978). Available alternative pest management strategies should also be calcu-

Table 2.   Chlorobenzilate use on citrus to control mites
(1975/76).

State	Crop	Amount (lb)	Acres treated	% of total acres
Florida	Orange	643,750	416,000	70
	Grapefruit	152,500	107,000	91
	Lemon	7,000	6,000	87
California	Orange	--	--	--
	Grapefruit	--	--	--
	Lemon	9,600	4,500	7
Texas	Orange	40,250	15,000	52
	Grapefruit	61,250	22,000	49
Arizona	Grapefruit	1,750	900	9
	Lemon	3,900	2,500	12
Specialty Citrus		76,000	47,000	--
TOTAL		996,000	620,900	

Source: Doane Agricultural Services, Inc. 1976; FA, 1977

lated into the benefit analysis. Epstein (1972) stated that
hazards from a chemical need not necessarily be accepted,
even when matching benefits appear high, if equally effective
but nonhazardous alternatives are available.

Benefit Analysis of Chlorobenzilate

In any specific case, a number of complicating factors
can influence risk/benefit analysis. This section describes
some of the complexities involved in analysis of a single
chemical, the pesticide chlorobenzilate. A miticide used primarily by citrus growers, chlorobenzilate was placed on the
RPAR list by the Environmental Protection Agency because it
is a moderate carcinogen in laboratory animals and has adverse
effects on the testes of male rats. Here we examine some of
the economic and biological information concerning chlorobenzilate use on citrus, data that must be included on the benefit
side of a risk/benefit analysis.

Current Use Patterns

About 995,700 lb, or 89% of all the chlorobenzilate used
in the United States, is applied to citrus acreage (Doane
Agricultural Service, Inc., 1976). Florida growers treat
about 70%, 91%, and 87% of their orange, grapefruit, and lemon
acreage, respectively, with chlorobenzilate or chlorobenzilate
combinations (Table 2). Florida contains most of the citrus
acreage in the United States (795,000 of an estimated 1,185,000
acres), and this state accounts for 81% of all chlorobenzilate
used on citrus. Texas, with relatively little citrus acreage,
uses 101,500 lb of the pesticide, on approximately one-half
of its citrus acreage. California and Arizona growers use
very small amounts of the chemical.

Chlorobenzilate is used primarily to control the citrus
rust mite, Phyllocoptruta oleivora. These mites feed on the
skin of the fruit and, depending on the level and the time of
damage, the rind becomes "bronzed" or "russeted" (McCoy and
Albrigo, 1975). Russeted and bronzed fruit at present can
not be marketed under U.S. No. 1 Bright grade, but it can be
sold as fresh fruit under U.S. No. 1 Bronze or Russet grades.
Ziegler and Wolfe (1975) state that although grades for fruit
are based on external appearance, representing the degree of
appeal to the eye of the consumer, all U.S. No. 1 grades must
meet the same requirements for maturity and internal quality.
However, Bright fruit usually commands a higher market price
than the other fresh-fruit market grades. Thus, citrus rust
mite populations are controlled to prevent grade-lowering "russeting" and "bronzing" on fruit shipped to the fresh market.

There are some specialty markets, however, that pay more for russeted fresh fruit. For example, Indian River grapefruit marketed in 1976/77 brought an average price of $5.72 per box for Bronze fruit, and only $3.45 per box for Bright fruit (Florida Fruit Digest, 1977), although the amount of Bronze fruit shipped totaled only 13,588 boxes. The reason for the price differential may be that russeted fruit is sometimes thought to be sweeter than bright fruit. The fruit may in fact be sweeter, because of water loss through the damaged skin and a· subsequent increase in sugar concentration in the fruit (Allen, 1979). The price difference quoted above is exceptional, but it does indicate that some consumers are aware of the often high internal quality of russeted fruit.

Since many growers do not want to be forced into the processed market by grade-lowering factors, they often treat for rust mite as insurance against this fate. Often, a grower can not afford to send fruit to the processed market if it had originally been tagged for the fresh-fruit market. Prices are much lower for processed fruit (Table 3). In California, after costs for picking, hauling, and handling are subtracted from the grower's price, a grower can actually lose money on processed fruit. Even in Florida and Texas, there are substantial differences in price between fresh and processed fruit.

Table 3. Fresh and processed fruit prices (1976/77).

State	Crop	Fresh ($)	Processed ($)
Florida	Orange	2.74	1.34
	Grapefruit	2.96	0.60
California	Orange	3.59	-0.60*
	Grapefruit	3.31	-0.47*
Texas	Orange	1.86	1.34
	Grapefruit	1.60	0.46

*Grower must pay picking, hauling and handling costs, and thus may receive less than the fruit cost to produce.

Source: FA, 1977.

Obviously, internal quality is of prime importance in fruit that is processed. One would expect citrus rust mite damage to be of little importance to the processor, as it affects primarily the rind of the fruit. Griffiths and Thompson (1953) went so far as to advocate a reduced spray program for this pest on fruit headed for the factory. However, as Ziegler and Wolfe (1975) noted, the capriciousness of processors toward external appearance makes the implementation of a reduced spray program difficult to achieve. For example, when yields are high and the fresh-fruit market is depressed, processors may discount fruit for external damage. During periods of high demand, processors may ignore external appearance and accept all fruit meeting internal quality standards.

The trend in the citrus-growing industry has been for increased production of fruit for processing, especially in Florida, where over 95% of the oranges are now processed (Huang and Duymovic, 1976; FA, 1977). Again, internal quality should be of prime importance to anyone buying fruit to process. To assess the benefits of chlorobenzilate use, then, one must ask whether the citrus rust mite reduces either internal quality or yields of fruit.

Damage Due to Citrus Rust Mite

Most of the early studies of citrus rust mite damage measured the impact of pest control measures on the basis of reduction in mite numbers, with little concern for the effect of control measures on crop yields. However, several studies that show the effects of damage on yield are available (Table 4). For oranges, the data indicate that yield remains the same in unsprayed and sprayed plots. In the study of McCoy et al. (1976), greasy spot disease, a fungus, was thought to have caused the lower yield. For grapefruit, a slight (about 5%) yield reduction due to rust mite damage may occur. However, this is a trend only, and not statistically significant, as grapefruit yields normally vary considerably among trees (Griffiths and Thompson, 1953).

Recent work has shown that the effects of citrus rust mite feeding are not serious until 50-75% of the surface of the fruit is russeted (Allen and Stamper, 1979). In this study, less than 5% of the oranges and 40% of the grapefruit in unsprayed groves was heavily scarred by rust mite. In another study, 16-38% of the fruit was russeted in unsprayed orange groves (McCoy, 1977). Also, damaging infestations usually do not last long in unsprayed groves in Florida, as natural enemies of the citrus rust mite control populations. At high mite densities and under normal weather conditions, a fungal disease of the mites, Hirsutella thompsonii, can reduce the mite population below economic levels during the summer months (McCoy et al., 1976). Indeed, the fungus was found to thrive in plots receiving no chlorobenzilate sprays to control the rust mite populations. Chlorobenzilate can reduce certain entomopathogenic fungi by 50-60% (Olmert and Kenneth, 1974).

Table 4. Yields of citrus in sprayed and unsprayed plots in Florida.

| Crop | Yield[a] | | Source |
	Spray	No spray	
Valencia oranges	2.9 boxes/tree	2.9 boxes/tree	Griffiths and Thompson, 1953
seedy grapefruit	10.0, of 13.4 expected ave. of boxes/tree	8.7, of 10.8 expected ave. of boxes/tree	Griffiths and Thompson, 1953
orange	201 boxes/acre	201 boxes/acre	Simanton, 1962
grapefruit	307 boxes/acre	245 boxes/acre[b]	Simanton, 1962
tangerine	310 boxes/acre	310 boxes/acre	Simanton, 1962
orange	3896 lb solids/acre	3500 lb solids/acre[c]	McCoy et al., 1976

a/ Reductions in yield from entire pest complex.

b/ Undetermined as to cause in yield reduction.

c/ Lower yield caused by greasy spot fungus.

In addition to discoloring the skin of citrus fruits, certain other detrimental effects have been attributed to the rust mite. For example, Allen (1978) found that the drop rate of fruit was affected by rust mite damage, although it did not increase above that of "bright" fruit until 75-80% of the surface skin was russeted. Allen (1979) also found that the weight of citrus fruit decreased with 50% or greater scarring, but the percentage of total soluble solids increased in proportion to the amount of rust mite damage. He also found that grapefruit achieved a smaller diameter when more than 87.5% of the surface was scarred. All of the above effects can most likely be attributed to increased water loss from scarred fruit.

In Florida, 78 and 72% of the orange and grapefruit acreage, respectively, is irrigated, while all of the bearing acreage in Texas and California is irrigated (FEA, 1976). Irrigation of citrus groves with citrus rust mite infestation would alleviate some of the problems of water loss caused by severe russeting. McCoy et al. (1976) stated that if 50% of the fruit in a grove is severely russeted, the fruit can be harvested early or irrigated, to save the crop. Also, heavy mite feeding on citrus leaves generally does not affect tree vigor, and can be offset by timely irrigation (McCoy, 1976).

Alternatives to Chlorobenzilate

Several alternatives are available to the use of chlorobenzilate. One of the best arguments for continued use of the chemical is that it is a relatively specific miticide, and is not detrimental to insect parasites of other citrus pests (Townsend, 1976; Fisher, 1977). The organophosphate alternatives, such as ethion, are nonspecific. Sulfur is also detrimental to many parasites. The use of petroleum sprays offers a good alternative (Jeppson et al., 1955; Townsend, 1976; McCoy et al., 1976). The use of the correct grade of oil, applied properly, will generally not cause phytotoxic effects (Simanton and Trammel, 1966; Riehl, 1969). When considering alternatives to a pesticide such as chlorobenzilate, a decision maker must determine how a particular pesticide operates within the entire pest control system. Restrictions on one pesticide may cause the use of greater amounts and possibly of more dangerous chemicals by the grower as an alternative to the original pesticide.

One alternative that should be considered is that of not spraying, or of reduced spraying. Griffiths and Thompson (1953) went so far as to state that fruit on unsprayed trees "will be as good as, if not better than, on trees receiving a complete spray program." For a reduced spray program to be possible for growers, the practices of fruit buyers would have to change. For processing, especially, buyers would have to be required to accept fruit based only on internal quality.

Benefit Analysis

In conclusion, a pesticide benefit analysis should include the following points:

(1) The role of the pesticide in increasing crop yield. Unfortunately, data explicitly concerned with yield are often scarce. Also, data of this sort are more difficult to assess when "quality" of appearance as well as yield are affected by the pest. However, a benefit analysis should make explicit the difference between improving yields and improving cosmetic appearance only. In some cases, such as fruit headed to the processor, costs associated with poor external appearance should not be assessed in the same way as costs associated with yield or internal quality reductions.

(2) The relationship of the target pest with the entire pest complex. A number of pests inhabit the citrus ecosystem. The adverse effects of some, such as the greasy spot fungal disease, can be easily confounded with those of the citrus rust mite.

(3) The effects of the pesticide on natural enemies. Some of the important pests of citrus, such as scale insects, are often under fairly effective biological control. Chlorobenzilate is a relatively specific miticide, while certain alternative chemicals may disrupt the biological control of the other pest species. However, chlorobenzilate may affect the natural control of the target species, the citrus rust mite, by interfering with control by the pathogenic fungus, Hirsutella.

(4) Availability of alternatives. In the citrus ecosystem, especially in Florida, oil can be used as an alternative pesticide that will not disrupt integrated pest management programs.

In sum, pesticides have economic benefits, but these should be kept in perspective. It is becoming increasingly important that benefits of pesticides be completely and fairly assessed. The "benefit" side is as important as the "risk" side of the equation in making policy decisions.

References

Allen, J.C. 1978. The effect of citrus rust mite damage on citrus fruit drop. J. Econ. Entomol. 71:746-750.

Allen, J.C. 1979. The effects of citrus rust mite damage on citrus fruit growth. J. Econ. Entomol. (in press).

Allen, J.C. and J.H. Stamper. 1979. The frequency distribution of citrus rust mite damage on citrus fruit. J. Econ. Entomol. (in press).

Borlaug, N.E. 1972. Mankind and civilization at another crossroads in balance with nature -- a biological myth. BioScience 1:41-43.

Doane Agricultural Service, Inc. 1976. Current pesticide use and user profiles for selected pesticide intensive crops, report no. 4 of "Pesticide use data on selected specialty crops", EPA contract 68-01-1928.

Durost, D.D. 1970. Crop production per acre and corn yield trends. 1970 Agr. Outlook Conf., Washington, D.C. Feb. 16-19.

Epstein, S.S. 1972. Information requirements for determining the benefit-risk spectrum. In Perspectives on Benefit-Risk Decision Making. Report by Comm. Pub. Eng. Pol., National Academy of Engineering. Apr. 26-27, 1971. Washington, D.C. 157 pp.

FA. 1977. Citrus Summary 1977. Florida Dept. Agr. Conserv. Serv., Florida Crop Livestock Rept. Serv., Orlando, Fla.

FCH. 1977. Farm Chemicals Handbook. Meister Publishing Co., Willoughby, Ohio.

FEA. 1976. Energy and U.S. Agriculture: 1974 Data Base. Federal Energy Administration, Office of Energy Conservation and Environment. Vol. I. FEA/D-76/459. Washington, D.C.

Fisher, J. 1977. Citrus integrated pest management program. The Citrus Industry. April: 9-10, 12, 15-16, 18-20.

Florida Fruit Digest. 1977. Prices compiled by the growers' administrative committee. Vol. 42. Florida Fruit Digest Company. Jacksonville, Florida.

Griffiths, J.T. and W.L. Thompson. 1953. Reduced spray
   programs for citrus for canning plants in Florida.
   J. Econ. Entomol. 46:930-936.

Hanley, J.W. 1979. The can-do spirit. Newsweek, Jan. 8.
   p. 7.

Headley, J.C. 1968. Estimating the productivity of agricultural pesticides. Am. J. Agr. Econ. 50:13-23.

Huang, B.E. and A.A. Duymovic. 1976. U.S. grapefruit trends
   and outlook. The Fruit Situation. TFS-198:46-52.

Hyslop, J.A. 1938. Losses occasioned by insects, mites,
   and ticks in the United States. E-444, USDA, Washington,
   D.C. 57 pp.

Jeppson, L.R., M.R. Jesser, and J.O. Complin. 1955. Control
   of mites on citrus with chlorobenzilate. J. Econ.
   Entomol. 48:375-377.

Jugenheimer, R.W. 1976. Corn: Improvement, Seed Production,
   and Uses. John Wiley and Sons, New York. 670 pp.

Lave, L.B. 1972. Risk, safety, and the role of government.
   In Perspectives on Benefit-Risk Decision Making. Report
   by Comm. Pub. Eng. Pol., National Academy of Engineering,
   Apr. 26-27, 1971. Washington, D.C. 157 pp.

Lupton, F.G.H. 1977. The plant breeders' contribution to
   the origin and solution of pest and disease problems.
   pp. 71-81 in Origins of Pest, Parasite, Disease and Weed
   Problems. J.M. Cherrett and G.R. Sagar, eds. Blackwell
   Scientific Publications, Oxford. 413 pp.

Marlatt, C.L. 1904. The annual loss occasioned by destruc-
   tive insects in the United States. pp. 461-474 in Yearbook of the Department of Agriculture. U.S. Govt.
   Print. Off., Washington, D.C.

McCoy, C.W. 1976. Leaf injury and defoliation caused by the
   citrus rust mite, Phyllocoptruta oleivora. Fla. Entomol.
   59:403-410.

McCoy, C.W. 1977. Resurgence of citrus rust mite populations following applications of methidathion. J. Econ.
   Entomol. 70:748-752.

McCoy, C.W. and L.G. Albrigo. 1975. Feeding injury to the
   orange caused by the citrus rust mite, Phyllocoptruta

oleivora (Prostigmata: Eriophyoidea). Ann. Entomol.
Soc. Am. 68:289-297.

McCoy, C.W., R.F. Brooks, J.C. Allen, A.G. Selhime, and
W.F. Wardowski. 1976. Effect of reduced pest control
programs on yield and quality of 'Valencia' orange. Proc.
Fla. State Hort. Soc. 89:74-77.

NAS. 1968a. Weed Control. Principles of Plant and Animal
Pest Control. Vol. 2. National Academy of Sciences,
Washington, D.C.

NAS. 1968b. Plant-Disease Development and Control. Principles of Plant and Animal Pest Control. Vol. 1.
National Academy of Sciences, Washington, D.C.

Newsom, L.D. 1978. Progress in integrated pest management
of soybean pests. pp. 157-180 in Pest Control Strategies.
E.H. Smith and D. Pimentel, eds. Academic Press, New
York. 334 pp.

Olmert, I. and R.G. Kenneth. 1974. Sensitivity of entomopathogenic fungi, Beauveria bassiana, Verticillium
leconii and Verticillium sp. to fungicides and insecticides. Environ. Entomol. 3:33-38.

Pimentel, D. 1973. Extent of pesticide use, food supply,
and pollution. J. N.Y. Entomol. Soc. 81:13-33.

Pimentel, D. 1976. World food crisis: energy and pests.
Bull. Entomol. Soc. Am. 22:20-26.

Pimentel, D. 1977. Ecological basis of insect pest, pathogen and weed problems. pp. 3-31 in Origins of Pest,
Parasite, Disease and Weed Problems. J.M. Cherrett
and G.R. Sagar, eds. Blackwell Scientific Publications,
Oxford. 413 pp.

Pimentel, D., C. Shoemaker, E.L. LaDue, R.B. Rovinsky, and
N.P. Russell. 1977a. Alternatives for reducing insecticides on cotton and corn: economic and environmental
impact. Report on Grant. No. R802518-02, EPA,
Washington, D.C. 147 pp.

Pimentel, D., E. Terhune, W. Dritschilo, D. Gallahan, N.
Kinner, D. Nafus, R. Peterson, N. Zareh, J. Misiti,
and O. Haber-Schaim. 1977b. Pesticides, insects in
foods, and cosmetic standards. BioScience 27:178-185.

Pimentel, D., J. Krummel, D. Gallahan, J. Hough, A. Merrill, I. Schreiner, P. Vittum, F. Koziol, E. Back, D. Yen, and S. Fiance. 1978. Benefits and costs of pesticide use in U.S. food production. BioScience 28;772, 778-784.

Pimentel, D., D. Andow, R. Dyson-Hudson, D. Gallahan, S. Jacobson, M. Irish, S. Kroop, A. Moss, I. Schreiner, M. Shepard, T. Thompson, and B. Vinzant. 1979a. Environmental and social costs of pesticides: a preliminary assessment. Manuscript.

Pimentel, D., J. Krummel, D. Gallahan, J. Hough, A. Merrill, I. Schreiner, P. Vittum, F. Koziol, E. Black, D. Yen, and S. Fiance. 1979b. Benefits of pesticide use in U.S. food production. In Pesticides: Role in Agriculture, Health, and Environment. T.J. Sheets and D. Pimentel, eds. Humana Press, Clifton, N.J. (in press).

PSAC. 1965. Restoring the Quality of Our Environment. Report of the Environmental Pollution Panel, President's Science Advisory Committee, The White House, Washington, D.C.

Riehl, L.A. 1969. Advances relevant to narrow-range spray oils for citrus pest control. Proc. 1st Internatl. Citrus Symp. 2:897-907.

Schalk, J.M. and R.H. Ratcliffe. 1976. Evaluation of ARS program on alternative methods of insect control: host plant resistance to insects. Bull. Entomol. Soc. Am. 22:7-10.

Simanton, W.A. 1962. Losses and production costs attributable to insects and related arthropods attacking citrus in Florida. USDA Coop. Econ. Insect Rept. 12: 1182.

Simanton, W.A. and K. Trammel. 1966. Recommended specifications for spray oils in Florida. Proc. Fla. State Hort. Soc. 79:26-30.

Sweetman, H.L. 1958. The Principles of Biological Control. Wm. C. Brown, Dubuque, Iowa. 560 pp.

Tihansky, D.P. and H.V. Kibby. 1974. A cost-risk-benefit analysis of toxic substances. J. Environ. Sys. 4:117-134.

Townsend, K.G. 1976. Two year summary of extension inte-
    grated pest management program. Proc. Fla. State Hort.
    Soc. 89:59-62.

USDA. 1936. Agricultural Statistics 1936. U.S. Department
    of Agriculture, U.S. Govt. Print. Off., Washington, D.C.

USDA. 1954. Losses in Agriculture. Agr. Res. Serv. 20-1.
    190 pp.

USDA. 1961. Agricultural Statistics 1961. U.S. Govt.
    Print. Off., Washington, D.C.

USDA. 1965. Losses in Agriculture. Agr. Handbook No. 291.
    Agr. Res. Serv., U.S. Govt. Print. Off., Washington,
    D.C.

USDA. 1966. Food. Consumption, prices, expenditures. Econ.
    Res. Serv., Agr. Econ. Rep. No. 138.

USDA. 1975a. Farmers' use of pesticides in 1971...extent of
    crop use. Econ. Res. Serv., Agr. Econ. Rep. No. 268.
    25 pp.

USDA. 1975b. Food. Consumption, prices, expenditures. Econ.
    Res. Serv., Suppl. Agr. Econ. Rep. No. 138, Washington,
    D.C.

USDA. 1978. Farmers' use of pesticides in 1976. E.S.C.S.
    Agr. Econ. Rep. No. 418.

van den Bosch, R. and P.S. Messenger. 1973. Biological
    Control. Intext Educational Publishers, New York.
    180 pp.

Woodwell, G.M. 1978. Paradigms lost. Bull. Ecol. Soc. Am.
    59:136-140.

Ziegler, L.W. and H.S. Wolfe. 1975. Citrus Growing in
    Florida. The University Presses of Florida, Gainesville.
    246 pp.

_____ *Jerry D. Stockdale*

6. Pest Management and the Social Environment: Conceptual Considerations

Abstract

   Approaches used in analyzing social and environmental
impacts of technological and policy change are summarized.
A framework is developed which emphasizes possible relationships between technological change and personal and social
well-being. Some implications of the framework for assessing pest management strategies are considered.

> "Would you tell me, please, which way I
>   ought to go from here?"
> "That depends a good deal on where
>   you want to get to."
>   Alice in Wonderland

Where We Want to Get to

   What are the most important likely impacts of pest
management technologies on persons and the social environment? How are these impacts to be assessed? Given alternative technologies, policies, and impacts, what criteria
should be used in selecting among alternative courses of
action?

   These seemingly straighforward questions are, in
reality, extremely complex. While some impacts of pest control practices spring readily to one's attention, e.g.,
hazards to health from pesticides, others are less immediately apparent yet potentially very important, e.g., impacts
on employment, population location, and economic concentration. Nor are various potential impacts equally easy to
measure. Even assuming perfect knowledge and quantification
of all likely impacts (an heroic assumption if there ever
was one), what criteria should be used in making policy

decisions? On what grounds are the impacts associated with a particular choice to be preferred over the impacts expected to accompany possible alternatives? What is the goal of all the analysis? Where "do we want to get to"?

In this chapter it is assumed that "where we want to get to" is some optimum level of quality of life or well-being for present and future generations.* Unfortunately, the meaning and sources of life quality are far from clear. Are we referring to objective conditions? What conditions? Or should we be more concerned with subjective assessments -- with how people feel? How realistic are various assumptions about connections between specific social and environmental impacts and life quality? If the osprey becomes extinct, does our life quality really change very much? How much? In what ways?

The assessment of social impacts of pest management technologies is made even more complex by the fact that such technologies are a subset of the more general phenomena of technological change in agriculture. Pest control techniques are usually components of "packages" of agricultural technology. And agricultural systems are, themselves, embedded in socio-cultural systems, which are also undergoing diverse forms of change. Most of the important questions to be considered in assessing impacts of pest management strategies and policies are, therefore, specific examples of more general issues in the analysis of societal and environmental change. It is, thus, not surprising that as one explores issues of assessing impacts of alternative pest management systems a variety of perspectives are available as possible guides for analysis. Some of these perspectives, including approaches to quality of life, are considered in the following section. Then a general framework for linking social, cultural, demographic, and environmental changes and quality of life issues is presented. The final section includes specific suggestions for assessing impacts of pest management alternatives.

The objective of this chapter, then, is to shed some light on the three questions which opened this section by developing and presenting a general framework for analysis of pest management technologies and policies and by exploring some implications of the framework. It is also intended that this framework should fill a major gap in the pest control literature. While social impacts have been considered in some recent sources (National Academy of Sciences, 1975;

---

* The terms well-being, quality of life, and life quality are used interchangeably in this chapter.

Smith and Pimentel, 1978), explicit statements of assumptions about expected relationships between social, environmental, and health impacts and personal well-being have been lacking. This has often resulted in failure to come to terms with the wide variety of ways in which any given technology or policy can influence the lives of people. Too little attention has been devoted to "where we want to get to."

Approaches to Impact Assessment

Major pest control strategies in use today include use of chemicals, biological control, genetic resistance, cultural practices and various combinations of these, plus other activities such as monitoring pest numbers. Each of these technologies is subject to rapid change, as is the context within which they are used. Pest control strategies are, thus, dynamic systems. This increases both the difficulty of impact assessment and its necessity.

Perspectives currently available for analysis of pest management alternatives are examined in this section. These can be classified into three general categories: 1. frameworks for analyzing systems, 2. perspectives which start with a particular kind of change and trace its impacts, and 3. perspectives which focus on desired outcomes, sometimes working back to possible sources. Each of these will be considered as background for the next section.

Systems Perspective

In the systems perspective, systems are seen as constituted of interrelated parts. Analysis of such systems usually involves analysis of the characteristics of the parts, the interrelations among them, and patterns of change over time. The most obvious systems perspective of consideration of environmental impacts of pest management alternatives is the ecosystem perspective. Writings in this perspective are especially helpful in pointing out the interconnectedness of physical, biological and social phenomena and for analyses of possible long-term environmental impacts of particular courses of action (Carson, 1962; Commoner, 1971; Ehrlich, 1968; Miller, 1975).

The literature in anthropology is generally characterized by systems perspectives and several anthropologists have incorporated environmental variables into analysis of sociocultural systems (Geertz, 1963; Rappaport, 1967; Vayda, 1969; Harris, 1974; and Bennett, 1976). Rappaport's analysis of the pig-ritual system among the Maring and Harris' analysis of India's sacred cows are useful in showing the complexity and importance of interactions among environmental, demographic, economic, social and cultural variables and the potential for unexpected negative consequences if such systems are disrupted.

In sociology the importance of relationships among social organization, cultural, demographic, environmental, and technology variables has been an important feature of some of the writings in the Human Ecology perspective, especially those of Hawley (1950) and Duncan (1959). The writings of Hawley and Duncan are important not only for the general perspective they present but also for directing the attention of (some) sociologists to environmental and demographic concerns. Their writings are, however, open to criticism for a de-emphasis of political-economic considerations. A sociologist who was sensitive to issues of the political-economy and to environmental concerns was the late Charles Anderson (1976). In economics Boulding (1973) has applied a systems perspective to environmental issues while in political science Ophuls (1977) has emphasized a socio-ecological perspective.

Other well-known examples of the systems perspective include the systems modeling efforts of the first and second Club of Rome reports by Meadows, et al. (1972) and Mesarovic and Pestel (1974). While these, especially the first, have been the subject of considerable debate, they do strongly call attention to issues of resource depletion and pollution and suggest possible long term interactions among demographic, economic, technological, environmental and quality of life variables. Certain other writings in the "growth-steady state" literature, for example those of Daly (1973), Ophuls (1977), Anderson (1976) and Heller (1975), place more emphasis on economic and political factors while considering issues of pollution and resource depletion.

This concern for relationships between economic, political, social, and environmental variables has long been evident in the literature on international development. Recent writings in this area, each with very different emphases and biases, include Mellor (1976), Paige (1975) and Perelman (1977). Some of the analyses of the social and environmental constraints and impacts of the Green Revolution also exhibit a systems perspective (Perelman, 1977).

Starting with Change and Searching for Impacts

While this approach has emerged from a variety of sources and concerns, one of its key elements today is cost-benefit analysis. Cost-benefit analysis, an outgrowth of

welfare economics, is specifically designed for use in public
decision making. Ideally, the goal is to discover the full
range of important benefits and costs of a particular technology, project or policy and to quantify them so the ratio
of costs and benefits can be compared in decision making.*
According to Mishan cost-benefit analysis

> ". . . implies a concept of social betterment that
> amounts to a <u>potential</u> Pareto Improvement . . .
> The project in question, to be considered as econ-
> omically feasible, must, that is, be capable of
> producing an excess of benefits such that every-
> one in society could, by a costless redistribution
> of the gains, be made better off." (Mishan,
> 1976, p. xii)

As Mishan suggests cost-benefit analysis in the public sector
is theoretically analogous to a costless program in the private sector which would internalize all external costs and
benefits.

    While comparison of costs and benefits in decision
making and policy setting makes undeniable good sense, it
is very complicated in practical application. The basic problem is one of identifying and valuing the full range of important costs and benefits. Given the difficulty of linking
particular secondary and tertiary benefits and costs to the
change in question, should they be included in the analysis?
Even if the decision is to include them, how should they be
valued? Can shadow pricing and compensation schemes, for
example, adequately represent the very real psychological
costs borne by victims of change programs? How realistic
are attempts to attach values to clean air, pure streams,
good health, occupational choice, etc. To leave such important costs and benefits out of the analysis constitutes an
obvious bias. Yet to include them introduces an element of
arbitrariness which many analysts are likely to find objec-
tionable. These problems are compounded by the differential
incidence on population segments of benefits and costs. In
many cases it is important to know not simply what the benefits and costs will be but what benefits will be received by
whom and what costs will be borne by whom. Differing levels
of information, organization and political power among population segments further complicate the matter. These issues
suggest that, even when stated in technical terms, the
necessity for value choices is inherent in cost-benefit

---

* I say ideally because it seems clear that not everyone who
  uses cost-benefit analysis really wants to maximize the
  ratio of <u>social</u> benefits and costs.

analysis. ". . . both the theory and practice of cost-benefit analysis reflect an ultimately political component. . ."
(Schnaiberg and Meidinger, 1978, p. 11).

Cost-benefit analysis is also complicated by issues of
assessing opportunity costs and taking account of time differences in assessing costs and benefits (discounting).
Nevertheless cost-benefit analysis and related procedures
for assessing "cost effectiveness" are now in wide use in a
variety of policy and administrative activities in the federal government.

One of the earliest applications of elements of cost-
benefit analysis to assessment of effects of pesticide use
in agriculture was a 1967 report prepared by Headley and
Lewis (1967) for Resources for the Future. Soon after that,
use of cost-benefit analysis was greatly stimulated by passage of the National Environmental Policy Act (NEPA).

NEPA was an outgrowth of a wide variety of social, cultural and environmental concerns. By the late 1960's acceptance of and faith in an automatic and inevitable connection
between technology, economic growth, and human well-being
was broken. The meaning of "progress" was much debated.
Such books as Marcuse's One Dimensional Man (1964) and
Ellul's The Technological Society (1964) questioned, sometimes very strongly, the trends in technological and social
change. Relationships between science, technology, social
change and well-being were widely discussed. The variety of
concerns is indicated by the articles in Technology and Man's
Future (Teich, 1972). This book also includes a section on
technology assessment, a concept which first came to national
attention in 1967 when Congressman Daddario introduced a bill
calling for establishment of a Technology Assessment Board.
The purpose of the board was to encourage the identification
and assessment of effects and implications of applied research and technology (Teich, 1972). In a 1968 statement
Daddario suggested a series of steps for technology assessment, including: 1. identification of impacts, 2. establishing cause and effect relationships, 3. determination
of alternative methods to implement the program, 4. identification of alternative programs to achieve the same goal
and identification of impacts, 5. measurement and comparison
of sums of positive and negative impacts, and 6. presentation of findings (Daddario, 1972, p. 216).

While he suggested that technology assessment had been
going on for years on an ad hoc basis, Daddario specifically
mentioned Carson's Silent Spring (1962) as having provided
impetus for explicit and formalized analysis of impacts of

science and technology. According to Daddario,

> "Rachel Carson's, "Silent Spring" brought the reali-
> zation of how quickly we had accepted the pest
> control properties of certain chemicals without
> questioning what the consequences of their wide-
> spread dissemination might do to valuable insects,
> fish and wildlife." (1972, p. 207)

As was the case for technological and societal change in
general, applications of technology to agricultural production were, until recently, generally equated with progress.
New technology meant more food for less labor and release
from some of the most tedious and "backbreaking" work. The
forced migration of millions of people from employment on
farms was often viewed as a net benefit since this increased
the availability of workers for other productive activities.
In rural sociology an important subfield for research was
"adoption-diffusion" research with its implicit goal of
speeding rates of adoption of new technologies both domesti-
cally and in the less developed countries (See, for example,
Rogers, 1962, Diffusion of Innovations).

By the late 1960's the assumed connections between
changes in U.S. agriculture and progress were increasingly
the subject of debate. On the international scene the problematic aspects of development (modernization) were dramatically portrayed as early as 1959 in Achembe's novel on social
disruption in Africa and in 1962 in Nair's Blossoms in the
Dust--the Human Factor in Indian Development. A recent book
on this same theme is Bodley's Victims of Progress (1975),
in which progress is seen as a largely pernicious myth. The
technological miracles of the Green Revolution have also been
the source of a variety of questions about resource availability and costs, unemployment, population relocation, and
concentration of wealth and power.

As has already been indicated, one of the important
sources of debate in the United States was Silent Spring
(Carson, 1962). Another was the rediscovery of rural poverty
and service delivery problems in rural communities. By the
mid 1960's agricultural production and rural life were increasingly seen as components of complex ecological and
social systems and a variety of previously unrecognized problematic aspects of United States agricultural production had
become matters of public concern. To the pollution and
ecological impact issues raised by Carson, were added concerns about the rapid rates of consumption of non-renewable
resources, especially energy (Pimentel, et al., 1973;
Pimentel, et al., 1975; Steinhart and Steinhart, 1974).

Linkages between agricultural change and rural and urban
social impacts, having been suggested in The People Left Behind (National Advisory Commission on Rural Poverty, 1967)
were considered directly by Van Diver in "The Changing Realm
of King Cotton" (1966) and by Smith in his Studies of the
Great Rural Tap Roots of Urban Poverty in the United States
(1974). By 1976 the question was explicitly raised whether
the United States is developing ". . . a form of high energy,
high resource agriculture which will be impossible to sus-
tain once achieved" (Stockdale, 1977, p. 43). In 1978
Rodefeld, et al. published a lengthy volume on the causes,
nature, and consequences of changes in agricultural technology and alternative future courses of action (Rodefeld,
et al., 1978). Yet another critique of trends in agricultural technology and organization, this time from a Marxian perspective, is Farming for Profit in a Hungry World (Perelman,
1977).

A common idea in most of these sources is the recognition that agricultural technology tends to consist of packages of technologies rather than isolated technological innovations. In commercial agriculture, components of the
technological package for plant production include plant
breeding, pesticides, fertilizers, irrigation (in some
cases), and mechanization, along with compatible management
practices. Similar sets of technology could be suggested
for livestock production. Because of this it is difficult
and in some cases unrealistic to attempt to analyze possible
impacts of a particular category of technology in isolation.
While certain environmental impacts of pesticide use are
relatively easy to isolate, this is generally not the case
for social impacts. According to a recent National Academy
of Sciences report,

> "There is no satisfactory way to disentangle pest
> control in the corn/soybean sector from the com-
> plex of chemical and biological technologies that
> have provided so much of the upward thrust in agri-
> cultural productivity in the United States in re-
> cent decades. The core of technological change
> has been to substitute mechanical, chemical, and
> biological inputs for labor with land use re-
> maining approximately constant. These develop-
> ments both have been caused by and have contri-
> buted to the massive outflow of population from
> farms and rural areas. As these population shifts
> have occurred, rural people and rural institutions
> have been subjected to great stresses. . . .While
> no one could reasonably argue that pest control
> products or methods in corn and soybeans have

alone caused these changes, they are an important part of a combination of technologies that
contribute to the overall tendency of farms in
the Corn Belt to become larger in size and sales,
to employ less labor per unit of output, and to
become more specialized and technically efficient."
(National Academy of Sciences, 1975, p. 105)

It would be a serious mistake to fail to consider important social impacts simply because they arise from sets of
technologies rather than from any one component of a technological system. In some cases, however, causes of particular outcomes can be identified. Many of the social costs
of agricultural technology, for example, are more properly
attributable to mechanization than to chemicals, while many
negative environmental impacts are more attributable to
chemicals. In other cases, changes in one aspect of production, e.g., mechanization, are not possible without changes
in others, e.g., chemicals and plant breeding. Thus in some
cases the focus should be on particular technological components and their possible impacts. In others it is most enlightening to consider packages of technologies in assessing
possible impacts.

General categories of possible impacts of agricultural
technologies suggested by the literature include impacts on:
1. physical environment, including impacts on quality of
air, water, soil, fish and wildlife and availability and cost
of resources, 2. health and nutrition, including availability, quality, and cost of food and negative health impacts
from pollution, 3. social and demographic factors, including
farm size and concentration, capital requirements, employment opportunities on farms and the nature of farm work; occupational structure and employment opportunities in rural
communities, tax base, diversity of services and activities,
social and political inequality, and patterns of community
growth and decline; and location, diversity and differentia-
tion of occupations and employment alternatives at the societal level, population location and concentration, balance of
political power, level of living, and cultural differentia-
tion, and 4. psychic well-being, including satisfaction with
work, community and environment.

It is important to remember that technology assessment
in agriculture has been an aspect of (and is, thus, inextricably linked to) technology and impact assessment more
generally. In this larger arena a variety of environmental
and public interest groups were pushing for greatly increased attention by the federal government to potential negative environmental and social impacts and the National

Environmental Policy Act (NEPA) was enacted in 1969.

According to the legislation the purposes of NEPA are

". . . to declare a national policy which will encourage productive and enjoyable harmony between
man and his environment; to promote efforts which
will prevent or eliminate damage to the environment and biosphere and stimulate the health and
welfare of man; to enrich the understanding of the
ecological systems and natural resources important to the nation; and to establish a Council
on Environmental Quality." (Munn, 1975, p. 106)

In order to accomplish these goals Federal agencies are
expected to ". . . prepare an Environmental Impact Statement
(EIS) prior to taking any action or reporting on legislation
that would significantly affect the environment" (Munn,
1975, p. 106). According to Munn Environmental Impact Statements are to include:

"1.   Description of proposed action; statement of
      purposes; description of environment affected;
 2.   Relationship to land-use plans, policies, and
      controls for affected area;
 3.   Probable impact--positive and negative; second-
      ary or indirect; as well as primary and direct;
      international environmental implications;
 4.   Consideration of alternatives;
 5.   Probable adverse effects which <u>cannot be avoid-
      ed</u>;
 6.   Relationship between local and short-term uses
      and long-term environmental considerations;
 7.   Irreversible and irretrievable commitment of
      resources;
 8.   Description of what other Federal considera-
      tions offset adverse environmental effects of
      proposed action and relation of these to
      alternatives."
      (Munn, 1975, p. 107)

More recent policy statements on EIS by the Council
on Environmental Quality and the Environmental Protection
Agency are included in the appendices of the <u>Environmental
Assessment and Impact Statement Handbook</u> by Cheremisinoff
and Morresi (1977). SCOPE Report 5 from the International
Council of Scientific Unions Scientific Committee on Problems of the Environment (Munn, 1975) considers <u>environmental
impact assessment</u> from an international perspective.

In both books the environment is broadly interpreted to include aesthetic and social as well as physical and ecological concerns. Four general categories of environmental impacts suggested by Cheremisinoff and Morresi (1977) include: ecological, physical/chemical, aesthetic, and social.* SCOPE Report 5 (Munn, 1975) suggests fourteen general "Areas of Human Concern" as possible "impact categories." These include: 1. economic and occupational status, 2. social pattern or life style, 3. social amenities and relationships, 4. psychological features, 5. physical amenities (intellectual, cultural, aesthetic and sensual), 6. health, 7. personal security, 8. religion and traditional belief, 9. technology, 10. cultural, 11. political, 12. legal, 13. aesthetic, and 14. statutory laws and acts.

Given the general agreement on the importance of considering social as well as physical and ecological impacts of various kinds of changes plus the difficulty of assessing such impacts, it is not surprising that <u>social impact analysis</u> has emerged as both an aspect of environmental assessment and also as a separate category of analysis. While concern for developing social impact statements (SIS) flowered after the NEPA legislation, concern for assessment of social impacts was already advertised by the Federal Highway Administration as a top research priority as early as 1966 (Wolf, 1974). Broadly interpreted social impact assessment is much older than that. According to Wolf (1974, p. 2), "It is at least arguable that "social impact assessment" is what social science is all about, and always has been." But the term social impact assessment is a recent one and the well-formed theories and methods which the term implies are only beginning to be developed.

Approaches Which Move from Impacts Back to Sources of Change

This third and final category of approaches to analysis of impacts of change is concerned with "where we want to get to" -- with the goals of change, with outcomes to be sought and avoided. As has been the case for the other categories, only some of the research and conceptual work in this area can be summarized here. The goal is, again, to provide an overview of work from this perspective and to emphasize those studies most directly useful for assessing impacts of pest management strategies and policies. The references cited here vary in the extent of their explicit concern for relating the impacts or conditions studied back to their causes. Some have been primarily concerned with specifying what constitutes a high standard of living, personal and

---

* Cheremisinoff and Morresi cited Dee, N. <u>et al.</u> (1974).

social well-being and the good life, the good community, or
the good society. Other studies have emphasized the need for
theories concerned with connections back to antecedent conditions and changes. Some studies have been concerned with
evaluating a wide range of aspects of personal and social
well-being. Others have emphasized more limited impacts.
Epidemiological studies, analyzing the incidence of cancer
and other health problems across geographic space, work situ-
ations, and population segments and seeking possible causal
factors, constitute an example of this more specific approach. Another example is studies which begin with a particular negatively evaluated environmental change, e.g., change
in the quality of fish in a lake, and search for causes.
While such studies are indeed important, they are touched
on only indirectly here. In what follows the focus is more
broadly on the question of "where we want to get to" -- on
what constitutes personal and social well-being or quality
of life. Concepts to be considered here include social and
environmental conditions, standard of living, way of life,
quality of life and social indicators. Linkages back to
technological, policy and other types of change will be considered in the next section of the paper.

The experience of material super-abundance for more than
a very small segment of the population of any country is a
very recent phenomenon. Even today adequacy of food, shelter
and health care is a serious problem for much of the world's
population. It is not surpising, therefore that material
conditions and indicators of ability to satisfy basic physiological needs have been very important to assessments of personal well-being.

Direct measures of (material) level or standard of
living were developed by economists and sociologists in the
1930's and '40's in the form of "level of living indexes."
A 1948 rural sociology textbook by Landis contains a map
depicting level of living index values for all the counties
of the United States in 1940. According to the text the
following were taken into account in the calculation of scale
values: "percent with gross income over $600 and percent
with 1936 or later cars," "dwellings with running water and
percent with mechanical refrigeration," "median years of
schooling of persons 25 years of age and over," "radios in
dwelling, and rooms per person in occupied dwellings"
(Landis, 1948, p. 89). Other items often used in level of
living indexes included home furnishings, indoor plumbing,
central heating, electricity, telephone, and newspaper and
magazine subscriptions.

While such material considerations still represent the good life to many of the world's people, by the 1960's scientists and policy makers in the developed countries were increasingly turning their attention to other quality of life concerns. According to Parke and Seidman,

"In the 20 years between 1949 and 1969, median family income in the United States doubled and the Gross National Product more than doubled (in constant dollars). Yet the same period was marked by crime, drugs, racial unrest, demonstrations, and urban crisis, the discovery of environmental degradation, and the rediscovery of poverty in America. These events prompted doubts about the easy equation of economic growth and social progress and a widening sense that economic indicators alone no longer sufficed to measure that progress. Among social scientists and public administrators there arose a renewed interest in social measurement more broadly conceived. In the mid-1960's this interest emerged in the writings and speeches of social scientists, social commentators, and policymakers in the form of calls for the development of "social indicators," "social accounting," "social reporting," "measuring the quality of life," and "monitoring social change." (Parke and Seidman, 1978, p. 2)

Important early writings in this vein included: Technology and the American Economy (President's Commission on Technology, Automation, and Economic Progress, 1966) which called for the creation of a system of social accounts for use in measuring and assessing social changes; Social Indicators (Bauer, 1966) which originated in a concern for analysis of impacts of the space program on American society; two volumes of The Annals entitled Social Goals and Indicators for American Society (Gross, 1967a and 1967b), concerned with conceptualization and measurement of social indicators; and Toward a Social Report (U.S. Department of Health, Education and Welfare, 1970) in which statistics were presented and patterns of change were discussed for selected topic areas.*

In the early 1970's the literature on social indicators and quality of life increased greatly. The categories of social indicators mentioned most frequently in this

---

* Summaries of the early history of the social indicators movement can be found in Land (1975), Sheldon and Parke (1975), Liu (1976), and Parke and Seidman (1978).

Table 1. Three approaches to conceptualizing categories of well-being.

Category	Approaches		
	Conditions conducive to well-being	Objective indicators of well-being	Subjective indicators of well-being
Education	Availability of educational opportunities-- e.g., teacher-student ratios, variety of offerings	Educational attainment --e.g., years of school completed	Satisfaction with education
Employment	Availability of employment opportunities --e.g., numbers of jobs of various kinds	Employment activity --e.g., unemployment rates	Satisfaction with work
Housing	Availability of Housing --e.g., number of housing units of various types	Quality of housing lived in--e.g., persons per room	Satisfaction with housing
Health	Availability of Health Care--e.g., doctor-patient ratios, hospital beds per 1,000 population	Health status-- e.g., rates of infant mortality, life expectancy	Satisfaction with own health and with availability of services

literature can be summarized as follows (the number of men-
tions in each category is indicated in parentheses; the
studies reviewed are indicated after the listing of categories of indicators): income, income and property, standard
of living, consumption (17); leisure, everyday life and
leisure, recreation (16); education, opportunities for learning (14); employment, jobs, work, work relations (13);
housing, home, quality of home environment (13); health,
health and nutrition, health services (12); security, safety,
law and justice, social disorganization (11); the environment, natural environment, environmental quality (9);
family, family life, marriage (8); social mobility, social
opportunity and participation (6); transportation (5);
level of services (5); community, city, town (4); neighbor-
hood (4); friends, friendships, interpersonal relations,
social acceptance (4); national government (4) (Andrews and
Withey, 1976; Bestuzhev-Lada, 1978; Bharadwaj and Wilkening,
1977; Campbell, Converse and Rodgers, 1976; Christian, 1974;
Dillman and Tremblay, 1977; Fitzsimmons and Lavey, 1976;
Harwood, 1976; Johnston, 1977; Levy and Guttman, 1975;
Mangahas, 1977; Milbrath and Sahr, 1975; Schneider, 1975;
Tunstall, 1974; U.S. Department of Health, Education and
Welfare, 1970). For the most part these categories represent
judgements of researchers about what represents well-being
or is conducive to it. Of course not all of these categories
are equally important to well-being.

Three general approaches to social indicators are dis-
cernable in the literature: 1. measures of social and environmental conditions expected to be conducive to social
and personal well-being, 2. indicators purporting to measure
social well-being directly and objectively, and 3. subjective or perceptual measures of well-being. Table 1 suggests
that the way each of the categories of indicators is conceptualized varies according to the perspective of the researcher.

It is not unusual for a researcher to include indicators
from more than one of these approaches in a single study and
the need for both objective and subjective indicators is now
generally accepted. As Andrews suggests "Ideally there would
emerge two complementary sets of indicators--one consisting
of perceptual indicators, the other of "objective" indicators" (Andrews, 1974, p. 282). A report which emphasizes
both of the objective approaches but with some emphasis on
direct assessment of well-being is that of Bestuzhev-Lada
who prefers the term "way of life" to quality of life.
According to Bestuzhev-Lada his framework ". . . incorporates
both vital activity proper (including human behavior) and
partly its conditions, which are reflected in the categories

of level, quality and pattern of life" (1978, p. 9).

Undoubtedly the most comprehensive studies of subjective or perceptual indicators are The Quality of American Life by Campbell, Converse and Rodgers (1976) and Social Indicators of Well-Being by Andrews and Withey (1976). Both are concerned with isolating domains of subjective indicators and mapping their interrelations. Another study in this perspective is that of Levy and Guttman (1975). A very important conclusion from these, as well as from other studies of subjective indicators (Harwood, 1976; Bharadwaj and Wilkening, 1977; Wilkening and McGranahan, 1978), is the high degree of importance attached by study respondents to conditions central to their day-to-day living, including good health, happy marriage and family life, friends, good housing, an interesting job, and living in a place of one's choice. The importance of health is especially to be noted in the studies by Campbell, Converse and Rodgers (1976), Harwood (1976), Bharadwaj and Wilkening (1977) and Wilkening and McGranahan (1978). Harwood concluded that ". . . a health component is consistently of highest importance" (1976, p. 495). Wilkening and McGranahan concluded,

> "It appears that it is the unexpected and uncontrolled or stressful changes in status such as health, marital disruption and unemployment that affect life satisfaction the most. Positive as well as negative effects usually accompany changes in residence. (Wilkening and McGranahan, 1978, p. 218).

The literature on residential preferences is extensive. For our purposes it is sufficient to note that a variety of studies have indicated that residential location is important to people, that scores on subjective assessments of well-being tend to be higher in rural places than in large cities, and that residential preference studies have consistently shown small town and rural life to be strongly preferred by large numbers of Americans (Campbell, Converse and Rodgers, 1976; Dillman and Tremblay, 1977; Bharadwaj and Wilkening, 1977).* This is important, here, because one of the impacts of technological change in agriculture has been massive rural to urban migration (Stockdale, 1977).

Since the use of objective indicators is generally based on assumptions of direct or indirect connections to well-being and since subjective indicators are intended to

---

* Dillman and Tremblay (1977) contains a very brief but excellent summary of conclusions from studies of residential preferences.

also measure aspects of well-being, it is to be expected that
the two would be correlated in predictable ways. While this
is generally true, e.g., level of income is positively correlated with overall measures of well-being and unemployment
is associated with low scores, many of the relationships are
less strong than might be expected and some notable exceptions exist (Campbell, Converse and Rodgers, 1976, Schneider,
1975). Campbell, Converse and Rodgers found, for example,
that, while income is generally a good predictor of housing
quality, ". . . low income people are astonishingly satisfied
with their housing" (1976, p. 117). Cases such as this, plus
the fact that education tends to be negatively associated
with levels of satisfaction in some domains, suggests that
important variables are operating to mediate the impact of
objective conditions on subjective assessments. Possible
relationships between objective conditions and subjective
value context, including aspirations and expectations, and
subjective well-being are indicated in Figure 1. Similar
representations have been included elsewhere (Campbell,
Converse and Rodgers, 1976, p. 13; Stockdale, 1978) and the
importance of aspirations and expectations to subjective
assessments is now generally accepted.

Operating from the assumption that ". . . the components
of life satisfaction vary according to the interests, needs
and concerns of the person," Bharadwaj and Wilkening (1977,
p. 425) documented differences according to sex, age and
income in the importance of various satisfaction domains.
After analyzing relationships between political ideology
and satisfaction levels, Buttel, Wilkening and Martinson
concluded that "overall life satisfaction, . . . one of the
most prevalent indicators in well-being research. . ., shares
a substantial amount of common variance with political-
economic ideologies" (1977, p. 365). Thus the subjective
content of subjective indicators is indeed great.

While both objective and subjective indicators are
essential to assessments of social well-being, neither captures the essence of what quality of life or well-being
really is. Quality of life is not operationally defined by
such measures. It is something different and the difference
is important.

In this chapter it is assumed that quality of life
or well-being is ultimately based on the extent to which
physiological and psychological needs are met. Persons experience quality of life in the development of their poten-
tial--in self actualization. This is based on assumptions
that: physiological and psychological needs exist, they are
different from wants, they are limited, and they can

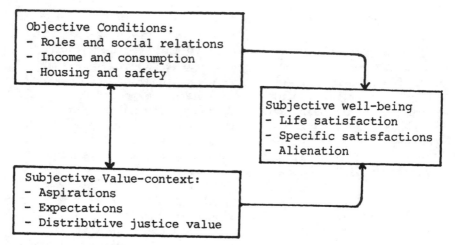

Figure 1. Domains of the life space. (Reprinted with permission from Social Indicator Models, edited by Kenneth Land and Seymour Spilerman, 1975. By the Russell Sage Foundation, New York.)

be satisfied.

Probably the perspective most relevant to this approach
is that of Maslow (1962) who suggested that it is essential
that two types of needs be met--basic needs and growth needs.
The basic (or deficiency) needs include physiological needs,
safety and security needs, love and belongingness needs,
and esteem needs. Maslow assumed that "frustration of basic
needs creates psychopathological symptoms, and their satisfaction leads to healthy personalities; both psychologically
and biologically" (Goble, 1970, pp. 50-51). Maslow referred
to the highest level of needs, the "self-actualization
needs" as growth needs. While not essential for health,
satisfaction of these growth needs was considered by Maslow
to be essential to maximum development of human potential--
to self actualization.*

Fromm has also assumed the existence of a level of
needs above purely physiological existence.

"The basic psychic needs stemming from the peculiar-
ities of human existence must be satisfied in one
form or another, unless man is to become insane,
just as his physiological needs must be met lest
he die." (Fromm, 1955, p. 67)

The existence of human needs is also assumed by Etzioni,

"There is a universal set of basic human needs which
have attributes of their own which are not determined
by the social structure, cultural patterns, or
socialization process." (Etzioni, 1968, p. 871,
emphasis added)

It is a basic assumption of this paper, then, that quality of life and well-being are attained when basic animal
and human needs are met and opportunities for individuals to
develop their potential are maximized--the more the opportunities to develop potential, the higher the level of well-
being. An alternative but related perspective is that of
McCall (1975) who specifies "general happiness requisites"
as "what it requires for an arbitrary member of the human
species to be happy" (p. 234) and suggests that "Quality of
life, as we shall define it . . . consists in the satisfaction of the general happiness requirements" (p. 235). McCall

---

* Sources on social indicators and quality of life in which
  Maslow's hierarchy of needs is discussed include:
  Stockdale (1973), McCall (1975), Liu (1976), and Campbell,
  Converse and Rodgers (1976).

agrees with Plato in equating the "happy life" with "the good or virtuous life" (p. 233).

A Framework for Assessing Impacts

Figure 2, a visual representation of a framework for assessing impacts of technological and policy change, is derived from the considerations of the previous sections of this paper. Figure 2 is intended to suggest the kinds of general and specific considerations and variables which must be included if any analysis of technological or policy change is to be complete.

Columns 1 and 2 are concerned with characteristics of socio-ecological systems and columns 3, 4 and 5 focus on individual level phenomena. For many purposes, of course, measures of individual level phenomena in columns 3 and 4 can be aggregated for populations. Column 1 of Figure 2 represents, in an abstract way, the overall socio-ecological system in which change takes place, columns 2, 3 and 5 represent specific categories of possible impacts and column 4 represents the ultimate concern of the analysis--the well-being of persons. It is assumed that quality of life (column 4) is different from but closely related to way of life (column 3) and subjective assessments of well-being (column 5).

The arrows suggest a linear flow from left to right. The idea is that a change in some feature of a socio-ecological system (column 1), e.g., technological change, will have a variety of impacts on other components of the system, resulting in changes in specific characteristics of socio-ecological systems (column 2), e.g., changes in availability, location, and type of employment opportunities, with these changes then influencing the way of life of at least some members of the population (column 3), e.g., place and kind of work and residential location, which will affect the extent to which various needs, e.g., belongingness and esteem needs, are met (column 4). The extent to which needs are met (column 4), plus relationships between persons' ways of life (column 3) and their cognitive characteristics, including aspirations, expectations, and values (column 3), will influence their subjective evaluations of well-being (column 5). While this linear flow of effects is useful for assessing impacts, it is necessarily an artificial abstraction; reality is characterized by much more complexity and interconnectedness than can be depicted here.

Column 1 in Figure 2 is very important because it calls attention to the fact that any technological or policy change occurs in a system of interconnected units and suggests the difficulty of assessing the full range of likely impacts of any given change.

Column 2 indicates, based on review of the literature, some important categories of relatively more specific characteristics of social systems and their environments which are susceptible to influence from changes in technology and policy and which are generally assumed to have important impacts on way of life and quality of life.

Way of life (column 3) is included in Figure 2 because it is assumed to be more directly connected to the extent to which needs are met and persons are able to develop their potential than either characteristics of social systems or subjective feelings of well-being. Way of life is concerned with the actual day-to-day lives of people, with the nature of their work, their leisure, their patterns of social interaction, their health, and their consumption--with the pattern, style, and level of their living. It is in day-to-day activities that persons' needs are met or not met and that they develop their potential. A complete assessment of impacts of technology must include, for example, not only impacts on rates of employment but also the quality and meaning of work experiences, not only impacts on the availability of health services but also impacts on the actual health of persons. Other aspects of way of life are indicated in column 3.

Column 4 is based on the assumption that, while way of life (column 3) and subjective indicators of well-being (column 5) are indicative of quality of life, neither captures the conceptual meaning of the term. Quality of life is assumed, instead, to be based on the extent to which needs are met and individuals have opportunities to develop their potential.

Research on subjective indicators of well-being has found several aspects or domains of life experience to be highly valued by respondents and to be very important to feelings of well-being. These are summarized in column 5. The emergence of health, friends, housing, work and community as very important is somewhat to be expected because of their relationships with physiological, belongingness, esteem, and self-actualization needs. This suggests that these domains should be emphasized in any analysis of impacts of technological or policy changes.

Figure 2. A general framework for assessing impacts of technological and policy changes.

(1)	(2)	(3)
Socio-Ecological System Level: Socio-Ecological System in Abstract	Socio-Ecological System Level: Selected Specific Characteristics of Socio-Ecological Systems ——▶	Individual Level: Personal Thought and Action

(1) Socio-Ecological System Level:
Socio-Ecological System in Abstract

*

SO = Social organization, nature of economic and political organization and organization of other institutional sectors

C = Aspects of culture other than technology, including symbols, knowledge, norms, beliefs, values

P = Population, demographic characteristics

T = Technology

E. = Physical evironment, including resource base

(2) Socio-Ecological System Level:
Selected Specific Characteristics of Socio-Ecological Systems

SELECTED CHARACTERISTICS OF SPECI-FIC SOCIAL SYSTEMS AND ENVIRON-MENTS, e.g., availability, quality, location of and differential access to:**

Employment
Education
Housing
Food
Health Care
Social services
Transportation
Other goods and services

plus:

Security and safety
Family life
Political participation
Consumption of resources
Environmental quality

(3) Individual Level:
Personal Thought and Action

WAY OF LIFE:
pattern, style & level of living, including:***

Economic activity
Social and political activity
Communication activity
Leisure, recreation & cultural activity
Transportation and travel
Residential location
Quality of physical environment
Quality of housing
Health
Level & type of education
Personal security
Family life

Cognitive characteristics, including beliefs, values, aspirations and expectations

Figure 2. (continued)

(4)
Individual Level: Meeting Needs and
Self-Actualization

(5)
Individual Level: Cognitive and Affective
Reactions

QUALITY OF LIFE:
(well-being) -- based on extent to which needs
are met and person is effective in developing
potential. Needs include:****

SATISFACTION AND FEELINGS OF WELL-BEING,
in general and for specific domains, including:*****

Physiological needs

Safety or security needs

Belongingness needs

Esteem needs

Self-actualization needs

Health   Family life
Work   Community
Housing   Food
Standard of living Education
Income and money Spiritual life
  matters   Sparetime activities
Friendships   Organizational in-
Natural environ-   volvement
  ment   National government

---

*   From section on "Systems Perspective," especially Duncan (1959).
**   From sections on "Starting with Change ..." and "Approaches which Move from Impacts ...."
***   Partially based on Bestuzhev-Lada (1978) and Podolakova (1978).
****   Listing of needs is from Maslow (1962).
*****   From section on "Approaches which Move from Impacts ...," especially Bharadwaj and Wilkening
        (1977) and Campbell, Converse and Rodgers (1976).

At least two general categories of needed theoretical and empirical analysis are suggested by Figure 2 and the preceding considerations. The first is analysis to increase our ability to describe, explain and predict relationships between selected changes in socio-ecological systems and the nature and distribution of other changes (impacts). This calls for greater understanding of the nature and dynamics of socio-ecological systems and subsystems. Special efforts are needed here because some of the connections between particular technological changes and impacts on some very important aspects of the day-to-day lives of people are difficult to establish. It would be a serious mistake to neglect some of the potentially most important impacts on the lives of people simply because of the difficulty of establishing connections back to a particular technological or policy change.

More research is also needed on relationships between quality of life as defined here and way of life, subjective indicators of well-being, and characteristics of socio-ecological systems. Such analysis should lead to greater understanding of human needs and potentials for self-actualization and of the importance of various personal activities and social and environmental conditions for meeting such needs. Even without further research, however, it is clear that the categories listed in columns 2, 3, and 5 in Figure 2 include potentially important categories of impacts of technological and policy changes.

Based on the content of Figure 2 and our literature review, it is possible to suggest some possible outcomes of pest management and other technological and policy changes which, were they to occur, could be expected to have negative effects on personal well-being. Specifically new technology or policy would reduce well-being if it were to: cause health problems; result in unemployment or a switch to a kind of employment considered less desirable; cause a change in residential location to a less desired place; negatively affect interpersonal relations, especially within the family; limit or reduce leisure time alternatives and restrict access to social, recreational and cultural activities; reduce personal security and feelings of safety; reduce satisfaction with the aesthetic quality of the natural environment; result in lower food quality or higher food prices; increase the cost of and reduce access to needed goods, such as fuel for heating homes; and increase inequalities in income, wealth, prestige, and power. Conversely, policies resulting in the opposite impacts would be ones which would increase overall levels of well-being. In reality, of course, impacts of any particular change will include both positive and negative effects.

The full extent of impacts of the outcomes just mentioned is much more difficult to assess than might at first be
imagined. The physical pain and discomfort experienced by
victims of health problems, for example, is only one of the
negative impacts associated with health problems. In an
assessment of the costs of cancer, Abt (1975) specified
eleven different categories of psychological costs, nine
categories of social costs and five categories of economic
costs. These various costs were then assigned to eight
different categories of people, including victims, spouses,
children, parents, siblings, friends, coworkers, and care
givers. While I find Abt's assignment of dollar values to
these costs both arbitrary and unrealistic, I do think the
attempt to specify the number of persons expected to bear
each of several specific kinds of psychic and social costs
to be very useful. Perhaps a general application of this
approach to assessment of pest management strategies and
policies and, indeed, to other types of technological and
policy change is in order. In such an approach, in addition
to assigning economic costs and benefits to the particular
alternatives under consideration, the most important likely
social and psychic impacts would also be specified and the
number of persons expected to experience these impacts would
be estimated. Judgments of the relative merit of the alternatives could then be made based on assessment of importance
of the various impacts to life quality and the number of
persons expected to experience each type of impact. A
further refinement, if desired, would be to subdivide the
summary according to the degree to which the respective
impacts would be experienced.

Implications for Assessment of Pest
Management / Strategies and Policies

As is the case for technological change and policy
generally, the likely impacts of pest management strategies
and policies are diverse, including some which are both obvious and relatively easy to measure and others which are
much less readily apparent and very difficult to measure.
Assessing impacts of pest management alternatives in agriculture is further complicated by the fact that pest management strategies, especially those including pesticides,
are usually components of sets of technologies, the impact
of any one aspect of which is difficult to identify. Nevertheless, it is important for policy purposes that assessment
be done and that the assessment be as complete as possible.

In such assessment it is essential that use of pest
management technologies be seen as occurring within socio-

ecological systems. This perspective directs attention to
both the diversity of kinds of possible impacts and the difficulty of predicting the full range of impacts which are
likely to occur.

Figure 2 and other analysis in this paper suggest that,
while it is often useful to start with a particular kind
of technology and work from there in an attempt to identify
important likely impacts, it is also possible to start with
the concept, quality of life, to delineate a variety of conditions and experiences considered to be conducive to life
quality and then ask which of these are likely to be influenced and in what ways by the pest management technology or
policy in question. The advantage of this latter approach is
that it suggests, even before the analysis begins, what some
of the most important categories of impacts might be. This
reduces the likelihood that important possible impacts will
be left out of the analysis and it focuses attention on the
purported goal of public policy--maximization of well-being.
The following is an example of how this approach might be
applied.

An Application of the Framework--Possible Impacts of Pest Insurance

Use of pesticides in corn production has a variety of
negative impacts, including those associated with toxic
buildup in the environment, damage to natural insect predators, and insect resistance. Nonetheless, because economic
costs of pesticides are low relative to possible costs from
pest damage, rates of pesticide use are high. Each year
approximately 50-60 percent of U.S. corn acreage is treated
with insecticides (National Academy of Sciences, 1975, p. 51).
It is estimated, however, that over half of this pesticide
use ". . . represents unneeded application--unneeded that is,
in hindsight" (National Academy of Sciences, 1975, p. 77).
What is happening is that many farmers are using pesticide
applications as a form of insurance against crop losses. One
way of reducing these "insurance" applications of pesticides
and, thus, avoiding some of the negative impacts, is by providing some other form of insurance or indemnification for
crop losses. Since "the major portion" of insecticides
used for corn are to control insects such as corn rootworms,
cutworms, wireworms and grubs (National Academy of Sciences,
1975, p. 51), an insurance program to reimburse farmers for
damage from these insects could (theoretically) greatly
reduce insecticide use.

Development of alternative insurance schemes and detailed analysis of likely impacts of such programs is beyond

the purposes of this chapter. Without specifying the characteristics of such a scheme, however, it is possible to use Figure 2 as a guide in suggesting some categories of possible impacts on quality of life.

In what follows it is assumed that an insurance scheme, somewhat similar to that used in hail insurance but otherwise unspecified, has been judged to be economically and politically feasible (not necessarily a realistic assumption). The items in columns 2, 3 and 5. are used as a guide for suggesting some important likely impacts of such a scheme on quality of life.

Impact on Employment. It is expected that an insurance scheme would result in only minimal changes in the nature and availability of farm work. Somewhat less time would be devoted to handling insecticides but the impact on most farms would be small. Nor is it expected that such schemes would materially affect farm numbers or opportunities for entry into farming. Such schemes would have more impact on employment in firms involved in farm supply and pesticide application, on chemical companies, on shippers of chemicals (including truck drivers), and, possibly, on firms providing inputs to chemical companies. The amount of the impact would depend on the effectiveness of the scheme in reducing insecticide use and on the economic alternatives available to the firms and individuals involved. Employment opportunities would be provided in the organizations created to develop, administer and carry out the insurance program. If the scheme were linked to a pest monitoring system, this would also create employment opportunities.

Impact on Income. Impact on income of farm families would depend on the nature of the scheme and on its effect on corn yields and prices. Participation in a voluntary program would be unlikely unless expected returns were at least approximately equal to those from insecticide use. If the program were to result in reduced crop yields due to increased pest damage, aggregate farm income for corn producers could actually increase due to price increases. Increased corn prices would probably have a negative impact on the incomes of some livestock producers. The income effect on farm supply firms, chemical companies, suppliers of inputs to chemical companies and transportation firms and employees would depend on the extent to which the insurance scheme was effective in reducing insecticide use and the extent to which suitable alternatives for use of capital and labor were available. It seems likely that farm supply firms and pesticide applicators would experience some reduction in income. Other income beneficiaries as a result of

the insurance scheme would be those involved in various
aspects of developing and carrying out the program.

Impact on the Cost and Supply of Food. If the insurance
scheme resulted in a reduction in total corn output, this
could result in increased domestic food prices and, thus,
have a negative impact on consumers. The amount of the
impact would depend, of course, on the extent to which
yields were reduced, plus market factors. It seems unlikely
that impacts on corn yields would be great enough to have an
impact on the cost and supply of food in the less developed
countries.

Impact on Cost and Supply of Energy. Any reduction in
use of pesticides would result in lower levels of consumption of petroleum and other resources in production and
distribution of pesticides. If corn yields were also reduced, less energy would be used in handling, drying, and
transporting the smaller crop. The total impact of these
effects, in comparison to total use of petroleum and other
resources, would be relatively small, however, and, thus,
impacts on prices would also be small.

Impact on Physical Environment. One of the most important impacts of an insurance program and, indeed, one of the
main reasons for developing such a program would be to reduce
the toxic load in soil, water and biological systems. Such
a scheme would positively affect the health of many organisms.

Impacts on Health. Reduced production and handling of
insecticides would reduce the chances of negative health
impacts for farmers and farm workers, pesticide applicators
and employees of farm supply firms, and chemical company
employees. Depending on the particular chemicals involved,
cutting use could also reduce possible health hazards for
the general public.

Impact on Family Life. The major impact on family life
would be for families of workers experiencing unemployment
and lacking suitable alternatives and for families of persons
finding employment in the insurance program. To the extent
that negative health impacts were reduced, this would also
have a positive impact on personal well-being and family
life.

Other Impacts. Some farmers are concerned about negative environmental and health impacts from pesticide use.
Implementation of a viable insurance scheme would provide
a more satisfying alternative for these farmers. While it

is difficult to assign values to this benefit, it would be possible, through survey research, to estimate how many farmers would experience such a benefit.

Categories of possible impacts from columns 2, 3 and 5 in Figure 2 which seem unlikely to be greatly affected by an insurance program include impacts on: education, housing, population location, community services, leisure and recreation activity, organizational involvement, and political participation.

The above listing of likely important impacts was prepared rapidly by examining the items listed in Figure 2 and asking whether each of them would be likely to be influenced by an insurance scheme. While not necessarily exhaustive of all possible impacts, the items above do indicate the kinds of considerations suggested by the framework. In order to make a more complete and detailed assessment of impacts, more information on the characteristics of the insurance scheme(s) and detailed information on various aspects of the existing situation, including environmental and health impacts of insecticides now in use, would be needed.

Summary

Several perspectives on the assessment of impacts of technological and policy changes have now been considered and a variety of sets of categories of possible impacts and indicators of well-being have been summarized. A framework for use in conceptualizing possible impacts has been presented (Figure 2), some of the most important categories of possible impacts have been delineated, and an example of how the framework might be applied has been presented. Important problems remain, including difficulties in quantifying impacts and in assessing the extent to which various indicators of impacts actually reflect quality of life. Also remaining are problems of establishing linkages between possible impacts and particular technological and policy changes.

References

Abt, C. 1975. The social costs of cancer. Soc. Indic. Res. 2:175-190.

Achembe, C. 1959. Things Fall Apart. Fawcett Publications, Greenwich, Connecticut. 192 pp.

Anderson, C. 1976. The Sociology of Survival. The Dorsey Press, Homewood, Illinois. 299 pp.

Andrews, F. 1974. Social indicators of perceived life quality. Soc. Indic. Res. 1:279-299.

Andrews, F. and S. Withey. 1976. Social Indicators of Well-Being. Plenum Press, New York. 455 pp.

Bauer, R., ed. 1966. Social Indicators. MIT Press, Cambridge, Mass. 357 pp.

Bennett, J. 1976. The Ecological Transition. Pergamon Press, New York. 378 pp.

Bestuzhev-Lada, I.V. 1978. Construction of a system of indicators of the level, quality, pattern, style and way of life of society: methodological problems. Unpublished paper. Soviet Sociological Association, Moscow.

Bharadwaj, L. and E.A. Wilkening. 1977. The prediction of perceived well-being. Soc. Indic. Res. 4:421-439.

Bodley, J.H. 1975. Victims of Progress. Cummings Publishing Company, Menlo Park, California. 200 pp.

Boulding, K.E. 1973. The economics of the coming spaceship earth. pp. 121-132 in Toward a Steady-State Economy. H.E. Daly, ed. W.H. Freeman and Company, San Francisco. 332 pp.

Buttel, F.H., E.A. Wilkening, and O. Martinson. 1977. Ideology and social indicators of the quality of life. Soc. Indic. Res. 4:353-369.

Carson, R. 1962. Silent Spring. Houghton Mifflin Company, Boston. 368 pp.

Campbell, A., P. Converse, and W. Rodgers. 1976. The Quality of American Life. Russell Sage Foundation, New York. 583 pp.

Cheremisinoff, P.N. and A. Morresi. 1977. Environmental Assessment and Impact Statement Handbook. Ann Arbor Science Publishers, Ann Arbor. 438 pp.

Christian, D. 1974. International social indicators: the OECD experience. Soc. Indic. Res. 1:169-186.

Commoner, B. 1971. The Closing Circle, Alfred A. Knopf, New York. 343 pp.

Daddario, E.Q. 1972. Technology assessment. pp. 201-219 in Technology and Man's Future. A.H. Teich, ed., St. Martins Press, New York. 274 pp.

Daly, H.E., ed. 1973. Toward a Steady-State Economy. W.H. Freeman and Company, San Francisco. 332 pp.

Dee, N., et al. 1974. Stratigraphy of the atlantic continental margin north of Cape Hatteras -- a brief review. USGS Open File Report, U.S. Department of the Interior, Washington, D.C.

Dillman, D.A. and K.R. Tremblay, Jr. 1977. The quality of life in rural America. The Annals, AAPSS 429:115-129.

Duncan, O.D. 1959. Human ecology and population studies. pp. 678-716 in The Study of Population. P.M. Hauser and O.D. Duncan, eds. The University of Chicago Press, Chicago. 864 pp.

Ehrlich, P. 1968. The Population Bomb. Ballantine Books, New York. 223 pp.

Ellul, J. 1964. The Technological Society. Alfred A. Knopf, New York. 463 pp.

Etzioni, A. 1968. Basic human needs, alienation and inauthenticity. Am. Sociol. Rev. 33:871.

Fitzsimmons, S.J. and W.G. Lavey. 1976. Social economic accounts system (SEAS): toward a comprehensive, community-level assessment procedure. Soc. Indic. Res. 2:389-452.

Fromm, E. 1955. The Sane Society. Holt, Rinehart and Winston, New York. 320 pp.

Geertz, C. 1963. Agricultural Involution, The Process of Ecological Change in Indonesia. University of California Press, Berkeley, California. 176 pp.

Goble, F. 1970. The Third Force. Grossman, New York.
201 pp.

Gross, B., ed. 1967a. Social Goals and Indicators for
American Society, Vol. I. The Annals, AAPSS, 371.

Gross, B., ed. 1967b. Social Goals and Indicators for
American Society, Vol. II. The Annals, AAPSS, 373.

Harris, M. 1974. Cows, Pigs, Wars and Witches. Random
House, New York. 276 pp.

Harwood, P. DeL. 1976. Quality of life: ascriptive and
testimonial conceptualizations. Soc. Indic. Res.
3:471-496.

Hawley, A. 1950. Human Ecology. Ronald Press, New York.
456 pp.

Headley, J.C. and J.N. Lewis. 1967. The Pesticide Problem:
An Economic Approach to Public Policy. Resources for
the Future, Inc., Distributed by the Johns Hopkins
Press, Baltimore, Maryland.

Heller, W.W. 1975. Coming to terms with growth and the
environment. pp. 208-228 in Population, Environment
and the Quality of Life. P.G. Marden and D. Hodgson,
eds., John Wiley, New York.

Johnston, D. 1977. Social Indicators, 1976. U.S. Department of Commerce, Washington, D.C. 564 pp.

Land, K. 1975. Social indicator models: an overview. In
Social Indicator Models. K.C. Land and S. Spilerman,
eds. Russell Sage Foundation, New York.

Landis, P. 1948. Rural Life in Process. 2nd edition.
McGraw Hill, New York. 538 pp.

Levy, S. and L. Guttman. 1975. On the multivariate
structure of well-being. Soc. Indic. Res. 2:361-388.

Liu, B.C. 1976. Quality of Life Indicators in U.S. Metro-
politan Areas, A Statistical Analysis. Praeger, New
York. 315 pp.

McCall, S. 1975. Quality of life. Soc. Indic. Res. 2:
229-248.

Mangahas, H. 1977. The Philippine social indicators project. Soc. Indic. Res. 4:67-96.

Marcuse, H. 1964. One Dimensional Man. Beacon Press, Boston. 260 pp.

Maslow, A.H. 1962. Toward a Psychology of Being. 2nd ed. Van Nostrand Reinhold, New York. 240 pp.

Meadows, D.H., D.L. Meadows, J. Randers, and W.W. Behrens III. 1972. The Limits to Growth. The New American Library, New York. 207 pp.

Mellor, J.W. 1976. The New Economics of Growth. Cornell University Press, Ithaca, New York. 334 pp.

Mesarovic, H. and E. Pestel. 1974. Mankind at the Turning Point. E.P. Dutton and Co., New York. 210 pp.

Milbrath, L.W. and R.C. Sahr. 1975. Perceptions of environmental quality. Soc. Indic. Res. 1:397-438.

Miller, G.T., Jr. 1975. Living in the Environment. Wadsworth, Belmont, California. 379, E161, A38 pp.

Mishan, E.J. 1976. Cost-Benefit Analysis. Praeger, New York. 454 pp.

Munn, R.E., ed. 1975. Environmental Impact Assessment: Principles and Procedures. SCOPE Report 5. International Council of Scientific Unions Scientific Committee on Problems of the Environment, available from John Wiley and Sons, New York.

Nair, K. 1962. Blossoms in the Dust -- the Human Factor in Indian Development. Praeger, New York. 201 pp.

National Academy of Sciences. 1975. Pest Control: An Assessment of Present and Alternative Technologies, Volume II, Corn/Soybeans Pest Control. NAS, Washington, D.C. 169 pp.

National Advisory Commission on Rural Poverty. 1967. The People Left Behind. U.S. Govt. Print. Off., Washington, D.C. 160 pp.

Ophuls, W. 1977. Ecology and the Politics of Scarcity. W.H. Freeman and Company, San Francisco. 303 pp.

Paige, J.M. 1975. Agrarian Revolution. The Free Press,
     New York. 435 pp.

Parke, R. and D. Seidman. 1978. Social indicators and
     social reporting. The Annals, AAPSS 435:1-22.

Perelman, M.M. 1977. Farming for Profit in a Hungry World.
     Allenheld, Osmun and Co., Montclair, New Jersey. 283 pp.

Pimentel, D., W. Dritschilo, J. Krummel, and J. Kutzman.
     1975. Energy and land constraints in food protein
     production. Science 190:754-761.

Pimentel, D., L.E. Hurd, A.C. Bellotti, M.J. Forster, I.N.
     Oka, O.D. Sholes, and R.J. Whitman. 1973. Food pro-
     duction and the energy crisis. Science 183:443-449.

Podolakova, K. 1978. The socialist way of life and the
     conditions of its formation. Unpublished paper. Re-
     Search Institute of Living Standard, Bratislava,
     Czechoslovakia.

President's Commission on Technology, Automation, and
     Economic Progress. 1966. Technology and the American
     Economy. U.S. Govt. Print. Off., Washington, D.C.

Rappaport, R. 1967. Pigs for the Ancestors, Ritual in the
     Ecology of a New Guinea People. Yale University Press,
     New Haven. 311 pp.

Rodefeld, R., J. Flora, D. Voth, I. Fujimoto, and J. Con-
     verse, eds. 1978. Change in Rural America, Causes,
     Consequences and Alternatives. The C.V. Mosby Company,
     Saint Louis, Missouri. 551 pp.

Rogers, E.M. 1962. The Diffusion of Innovations. The Free
     Press of Glencoe, New York. 367 pp.

Schnaiberg, A. and E. Meidinger. 1978. Social reality
     versus analytic mythology: social impact assessment of
     natural resource utilization. Unpublished paper. De-
     partment of Sociology, Northwestern University, Evanston,
     Illinois.

Schneider, M. 1975. The quality of life in large American
     cities: objective and subjective social indicators.
     Soc. Indic. Res. 1:495-509.

Sheldon, E.B. and R. Parke. 1975. Social indicators.
     Science 188:693-99.

Smith, E.H. and D. Pimentel, eds. 1978. Pest Control Strategies. Academic Press, New York. 334 pp.

Smith, T.L. 1974. Studies of the Great Rural Tap Roots of Urban Poverty in the United States. Carlton Press, New York. 144 pp.

Stockdale, J.D. 1973. Human potential: a perspective on poverty and quality of life. Growth and Change 4:24-28.

Stockdale, J.D. 1977. Technology and change in United States agriculture: model or warning? Sociol. Ruralis 17:43-58.

Stockdale, J.D. 1978. Quality of life in socio-ecological perspective. Unpublished paper. Department of Sociology, Anthropology and Social Work, University of Northern Iowa, Cedar Falls.

Steinhart, J.S. and C.E. Steinhart. 1974. Energy use in the U.S. food system. Science 184:307-316.

Teich, A.H., ed. 1972. Technology and Man's Future. St. Martins Press, New York. 274 pp. (A revised edition was published by St. Martins in 1977.)

Tunstall, D.B. 1974. Social Indicators, 1973. Office of Management and Budget, Washington, D.C.

U.S. Department of Health, Education and Welfare. 1970. Toward a Social Report. The University of Michigan Press, Ann Arbor. 101 pp.

Van Diver, J.S. 1966. The changing realm of king cotton. Transaction 4:24-30.

Vayda, A.P. 1969. Environment and Cultural Behavior. The Natural History Press, Garden City, New York. 485 pp.

Wilkening, E.A. and D. McGranahan. 1978. Correlates of subjective well-being in Northern Wisconsin. Soc. Indic. Res. 5:211-234.

Wolf, C.P. ed. 1974. Social Impact Assessment. Erda 5. Environmental Design Research Association. 198 pp.

7. Legal Aspects of Integrated Pest Management

Introduction

There is a growing consensus that American agriculture
relies too much on the application of chemical pesticides to
control pest damage. Rachel Carson's <u>Silent Spring</u> has made
society sensitive to many of the short- and long-run costs of
pesticides, but Daniel Zwerdling has recently charged that
"[t]he purpose of her widely acclaimed book -- to reverse the
tide of pesticide use -- has failed." (Zwerdling, 1977). The
way to maintain continued high crop production levels without reliance on repeated pesticide applications is said to be
Integrated Pest Management (IPM), which seeks to balance between chemical and nonchemical means of pest control to reduce pest damages to economically acceptable levels. Unfortunately, integrated pest management is not a uniform technology that can be imposed upon broad classes of pest managers
as we have imposed air and water pollution reduction technologies on industrial dischargers. The complex nature of IPM
therefore makes it difficult to answer the question: how can
the legal system contribute to the adoption of a pest control
strategy that decreases reliance on chemical pesticides and
still minimizes pest damage to valuable agricultural products?
This paper is a preliminary examination of this question.
First, the current laws regulating pesticide composition and
use are described in an effort to show the difficulty of in-
corporating IPM considerations into existing regulatory decisions. Second, a strategy appropriate to induce the
adoption of IPM is suggested. The aim of this paper is not
to present a comprehensive analysis of the legal system's
role in promoting IPM. My objective is only to raise the
consciousness of lawyers and nonlawyers about an important
but heretofore somewhat neglected area of pesticide regulation.

Background of Current Pesticide Laws

This country's high level of agricultural production depends on the constant application of progressively sophisticated technologies, especially chemical fertilizers and pesticides. New crop production technologies such as the no-till method continue the farmer's reliance on high levels of fertilizer and pesticide applications. Until the 1960s few questioned the benefits of this chemical technology. "Miracle" insecticides such as chlorinated hydrocarbons were assumed to be simply another example of the benefits that technology could produce for man. In the late 1950s and early 1960s, some scientists became concerned with two unanticipated potential social costs of pesticide use. First, pesticides may be chronically as well as acutely toxic to nontarget species, thus posing a risk of harm to man and his environment. Specifically, it is alleged that many pesticides are carcinogenic or mutagenic. Second, the continued use of pesticides may be an ineffective and increasingly costly way of controlling pests because many pests have become genetically resistant to the chemical and the chemical may kill the pest's natural enemies, causing target pest-resurgence. Toxicity and ineffectiveness are both important social problems. But, in the long run, the ineffectiveness of chemical pesticides may be the most important, for the consequence of increased pest resistance is that we must rethink the basic role of pesticide application in crop production strategies. This conclusion carries important consequences for the legal system, as the current federal and state laws regulating pesticide content and use are concerned almost exclusively with monitoring the composition and use of pesticides to insure that they are safe, rather than with the effectiveness of pesticides. The thesis of this paper is that current federal and state laws are based on an incomplete assumption about the social costs of pesticides. As a result, scarce regulatory resources are being devoted to the elimination of social costs that may be minimal compared to the social costs of applying ever more pesticides to achieve the same or decreased benefits. Farmers, the public, and the environment generally suffer when there is less bang for more bucks. Available regulatory resources should be at least partially reallocated to implement a regulatory program with the primary objective of reducing reliance on chemical pesticides rather than safety monitoring.

It will not be easy to change the current regulatory approach, for our current laws are a blend of the legacy of Rachel Carson's Silent Spring and contemporary philosophies

of technology assessment. The law of pesticide regulation
has passed through three stages of evolution as the relevant
class of persons, flora, and fauna deserving protection has
been broadened. From 1910 to 1947, a simple statute prohib-
ited the sale of adulterated or misbranded pesticides and
specified the required percentages of ingredients for the two
most common pesticides -- Paris green and lead arsenate
(Public Law 61-152, 1910). There was no comprehensive assessment at any time of the pesticide's impact, for the problem
was assumed to be deceit. Enforcement, if any, took place
only when a user reported an after-the-fact violation of the
Act. In 1947 Congress responded to the inadequacies of the
1910 law that became apparent due to the widespread domestic
use of synthetic organic pesticides. To provide the systematic screening of the new compounds needed to protect users,
the Federal Insecticide, Fungicide and Rodenticide Act of
1947 (U.S.C., 1947) was passed. Because the new pesticides
were more toxic and thus dangerous to valuable nontarget
species if not properly used, the pesticide problem was
assumed to be one of assuring users that the product was
efficacious and to provide adequate information to the user
to permit safe and effective choices. For the first time,
all pesticides had to be registered before they could be marketed and contain adequate warnings of the dangers of misuse
on the labels. In theory, FIFRA provided limited protection
from undesired side-effects to a rational user, but third
parties and the environment generally were given little protection from the long-term impacts of pesticide use. Safety
was initially narrowly defined and because of the provision
allowing a pesticide to be registered even if the USDA pro-
tested, the statute was administered as an automatic registration statute rather than a safety screening mechanism. As
public concern over DDT mounted, the legislation was amended
to expand somewhat the scope of protected interests. A product without an adequate label was classified misbranded, and
the concept of misbranding did permit some screening for
safety. The warning statement had to be adequate to prevent
injury to "living man and other vertebrate animals, vegetation
and <u>useful</u> invertebrate animals." (emphasis mine). In 1959
Congress provided that a pesticide could not be registered
regardless of label content if the pesticide was used as
directed or "in accordance with commonly recognized practice"
and still caused injury to the class of protected flora and
fauna (excluding weeds). This rather narrow vision of the
Garden of Eden could, nonetheless, have become the authority
for environmental screening.

Efforts to force USDA to ban DDT produced a series of
decisions that gave the Department of Agriculture (which was

charged with administering the statute) the discretion to ban pesticides for environmental reasons. This expanded reading of FIFRA became especially significant in 1970 when pesticide regulation was taken from the Department of Agriculture and given to the newly-created Environmental Protection Agency (EPA). EPA defined its mission solely as the promotion of environmental quality and took an active role in pesticide screening. The outer limits of FIFRA did not, however, have to be defined because in 1972 the 1947 legislation was over-hauled, the judicial constructions of the 1947 legislation validated, and the information disclosure registration provisions were supplemented with a regulatory mandate designed to subject all pesticides to rigorous screening to determine if they were unsafe, broadly defined, and thus compatible with environmental quality.

Current Pesticide Laws

During the public and Congressional debates over the new legislation, there was recognition that there were two problems with pesticide use: environmental safety and long-run ineffectiveness. Some members of Congress such as Senator Nelson of Wisconsin wished to address both problems. However, for two principal reasons Congress concentrated on the question of environmental safety. First, the "big" public issue was DDT. Public interest lobbying groups concentrated their efforts on the clean and quick solution of banning DDT and similar compounds rather than on a more comprehensive but messy attack on the general problem of pesticide use patterns. Second, environmentalists had to do battle with agriculturalists who were already unhappy with the transfer of pesticide regulation from the familiar recessed covers of the Department of Agriculture to the unknown and exposed shores of EPA, even though many of the technical personnel in USDA sailed with the transfer. Environmental regulation was bad enough, but the outright possibility of reducing the total amount of all pesticides used rather than banning one and substituting another was worse. Thus, environmental and agricultural personnel agreed to concentrate on safety rather than use reduction questions. The legislative history makes it clear, however, that some attention was given to the ineffectiveness of present use levels, for 1972 legislation does address the problem of use reduction indirectly. The next section discusses briefly the purpose and structure of the 1972 legislation with emphasis on the less emphasized and attenuated relationship between environmental screening and overall use reduction.

The 1972 legislation, the Federal Environmental Pesticide

Control Act (FEPCA)(U.S.C., 1978a), was shaped by the DDT experience and specifically by the Court's construction of FIFRA (D.C. Cir., 1971, 1976a). The DDT experience continues to dominate the EPA's pesticide policy followed in implementing the Act. In the 1950s, some scientists became concerned with the long-term impact of DDT on the environment due to residue concentration in food chains, but at the time concern over DDT was growing, it was not clear that harm to the environment generally made a pesticide unsafe. This fascinating, profound philosophical and perhaps unanswerable question was largely mooted when evidence linking DDT build-ups to cancer in humans appeared. The opponents of DDT were able to shift to a well-established and widely acceptable ground of attack, arguing that a chemical should be banned when it was likely to cause a risk of unintended harm to nonconsenting humans. The difficult legal issue was whether it was possible to ban a pesticide if the evidence showed only that there was a risk of harm to humans rather than cause-in-fact as lawyers defined the term. This risk of harm to humans could be based on inferences from tests performed on laboratory animals. The technical issue that had to be resolved, once the courts and Congress indicated, out of necessity, that proof of some risk level would be accepted for proof of cause in fact, was whether it was legitimate to infer a risk to humans solely from evidence that the chemical caused cancer in laboratory animals. A series of important judicial precedents interpreting both the 1947 and 1972 legislation established the legitimacy of risk as a basis for a decision, and that animal test results could serve as the basis for an inference of risk.

FEPCA is technically a law that supplements the label information disclosure requirements of FIFRA, with a regulatory structure designed to provide for the screening and continual evaluation of the safety of all pesticides. The two most important innovations of the 1972 legislation are that (1) all pesticides must be screened on a case-by-case basis for environmental safety, and then subjected to a benefit-cost analysis to determine whether they can be marketed, and (2) the use of the pesticide as well as the safety of the compound is evaluated. The crucial registration criterion is that the pesticide must perform its intended function and when used "in accordance with widespread and common practices," it does not cause "unreasonable adverse effects on the environment." (U.S.C., 1978b). The Act allows a pesticide to be screened at any one of three stages. In order to gain access to the market, a pesticide must be registered. Once it is on the market, it may be suspended pending a full cancellation hearing. Suspension is a quasi-summary procedure that removes a product from the market to avoid substantial harm to the public, pending a final decision about its safety. A cancellation is a permanent termination of a registration that bars the product, or perhaps only a use of the product, from the market. The standards for registration and cancellation are essentially the same; due to its emergency nature, the standard for suspension is stricter, as an "imminent hazard" must be established. To allow the EPA to recognize local and regional variations in need of a pesticide and the degree of risk exposure that results from a use, the EPA may register a pesticide for either general or restricted use, as well as cancel uses on a selective geographical basis. The former category, which was introduced into pesticide regulation for the first time in 1972, allows the EPA to specify the conditions, e.g., required equipment and degree of applicator training, under which a pesticide may be used.

FEPCA is a modern example of technology assessment legislation, which illustrates both the strengths and weaknesses of current approaches to the art of technology evaluation. The history of twentieth century regulation of business activity is largely one of the rejection of common law methods of assessing technology. In the 19th century the common law basically assessed technology after the fact, as the application of technology was presumed to be beneficial. Those who proved injury to a legally protected interest received compensation and sometimes could prohibit an activity. The ad hoc nature of after-the-fact assessment and the strictness of the standards of injury and cause incorporated into the common law have been gradually rejected as inadequate in a number of fields of activity, from the issuance of securities to the marketing of pesticides. Regulation was seen as the answer to the common law's limitations. The public interest would be protected through the consistent application of expertise. One theory of regulation urges the substitution of advance rules for damage actions to insure that activities would be carried out in a socially acceptable manner. This theory was first applied to pesticides in 1947 and has been carried over to the 1972 legislation.

FEPCA is a product of the environmental movement that reflects a growing skepticism about the consequences of the full force of the presumption that technology is beneficial and the corollary that any problems caused by a "bad" technology can be solved by a "good" technology. Ecology yields no moral imperatives, but it provides a counter presumption of caution based on the assertion that nature knows best. The environmental laws passed in the last decade are schizophrenic with respect to technology and, on the whole, represent only a modest weakening of the presumption that technology is beneficial. For example, The Clean Air and Water Acts of 1977 are classic examples of the theory that bad can be cured by good technology. They both seek to reduce the use of air and watersheds as sinks by forcing the application of higher and higher levels of technology on dischargers. Pesticide laws along with other laws regulating toxic substances incorporate the recently-recognized counter presumption of caution to some extent, for they admit the possibility that some technologies must be barred from the market or their use substantially restricted. Regulators are given the discretion to set conservative risk thresholds and to bar substances from the market after a benefit-risk analysis that weighs heavily low-level, low-probability risks.

This regulatory technique is of limited utility in addressing the problem of excessive pesticide use for two reasons. First, the assessment procedure is costly and lengthy, and thus only a few compounds can be screened at a time. Delay and regulatory inefficiency continue to plague the EPA (Aidala, 1978). Second, once an ingredient is screened and a decision is made to allow the pesticide on the market, unlimited amounts of the pesticide can be marketed, for it is presumed safe so long as the product is used consistent with the registration. Under a regulatory program designed to screen unsafe chemicals, the effectiveness as distinguished from the inefficacy of a pesticide, becomes irrelevant.

The important point here is that our current approaches to technology assessment generally stop short of requiring a fundamental reevaluation about resource allocation choices that gave rise to the need for regulation. Generally our laws, which, of course, are only reflections of societal attitudes, do not question the rationality of the conditions that produce problem-causing resource allocation choices. Instead, the undesired side-effects of a resource use choice are seen as costs to be minimized within the constraint that the underlying conditions cannot be questioned. This observation is not at all meant to advocate Luddite solutions or a naive, elegiac search for some simple but unattainable Eden. My argument is only that current laws designed to assess technology address a significant but ultimately narrow range of issues and further, that in some cases harder and deeper questions must be addressed, for conventional assumptions do not adequately reveal the hard choices society may be forced to make.

FEPCA therefore does not change the underlying theory of technology assessment adopted in 1974. It merely adds new and more stringent assessment standards. The limited modification of existing assumptions can be seen in the contrast between FEPCA and the Clean Air and Water Acts, which "force" the adoption of new control technologies. Not only must an activity be screened in advance, but the method of achieving the objective must also be specified, because Congress has set high -- some say unrealistic -- environmental quality goals. The desire to reach these goals has, of necessity, led to the adoption of "technology-forcing" legislation. One undertaking an activity cannot defend against the imposition of safety standards by proving that he conforms to the generally accepted state of the art if a more efficacious technology exists or will be developed in the foreseeable future. The defenses shift to issues of feasibility and cost. The concept of technology-forcing has obvious implications for the question of integrated pest management, which will be discussed subsequently.

FEPCA, although a limited modification of traditional assumptions about the desirability of chemical technologies, has the effect of putting the Environmental Protection Agency in a poor position to consider the problem of long-run pesticide ineffectiveness. The dynamics of the demand for pesticide regulation has led EPA to set very conservative risk assessment standards. Professor Edmund Kitch of the University of Chicago has described one consequence of this choice with respect to the activities of the Federal Food and Drug Administration. The EPA has more flexibility than the FDA, because EPA can condition a pesticide based on its use, as well as ban it, but the essential criticism of the FDA applies with equal force to the EPA:

> One of the important criticisms of the regulatory scheme that has been made is that it confronts the FDA with an all or nothing choice. Either a drug is approved for general marketing, subject only to the constraints of the label limitations, or it is not approved for marketing. And once it is approved for marketing, the formal regulatory review of safety and efficacy ends -- at the very time when the commercial sales, much higher in volume than experimental use and production could ever be -- are generating much more information of potential regulatory value. Since the FDA has an all or nothing choice, it tends to be very cautious before it says yes (Kitch, 1978).

EPA's conservatism is reflected in the regulations it published in 1976 to screen chemicals. Based on its success in establishing laboratory animal studies as a basis for inferring that a chemical exposes humans to a risk of cancer, the EPA created a procedure based on the low threshold carcinogenesis triggers. If an acute or chronic toxicity trigger is tripped, a rebuttable presumption against registration arises against the pesticide (Code of Federal Register, 1978). Once a chemical is "RPARed" the registrant bears the burden of rebutting the presumption. This can be done either by showing that the EPA's data alleged to trip the trigger was wrong, or by convincing EPA that the benefits of some or all uses outweigh the risks. The triggers are very conservative for benign as well as malignant tumors in the sensitive laboratory animal, the mouse, are sufficient evidence to issue an RPAR. Congress is not entirely satisfied with the RPAR regulations, which are of doubtful legality, and in 1978, it directed EPA to be more balanced. The wisdom and operation of the RPAR procedure is a separate subject worthy of a session. My discussion of the RPAR procedure is limited to illustrating the conservative nature of the EPA safety standards and to illustrate how the initial success in banning DDT has continued to shape the EPA's regulatory policy. The agency is now engaged in the task of hunting down the carcinogens and mutagens. No one doubts that there is a need to screen compounds for these effects, but one must ask at what costs and with what benefits. Given the difficulty in identifying causes of cancer and the remoteness of the exposure of most people to pesticides, it is arguable that the EPA is carrying a very costly program that yields few benefits to society and fails to reduce our reliance on chemical pesticides, thus not "solving" anything.

In December of 1978, EPA proposed replacing the compound-by-compound approach to pesticide registration with a generic approach, which will focus on the active ingredient and all the formulations in which it appears. Generic standards may speed the registration process, but it will not change the standards EPA is applying or the basic approach to pesticide regulation. A related consequence, which bears directly on the fate of integrated pest management, is that the adoption of a conservative strategy increasingly forces the EPA to make Hobsons' choices. These choices serve only to undermine the effectiveness of the agency's mission. Almost all important pesticides are potential RPAR candidates and so the agency finds itself making harder and harder choices, which are not politically popular among the industry, state pesticide regulators or influential members of Congress. The first regulatory decisions are always the easiest because substitute chemicals are available, but as more and more widely used pesticides are knocked out, the range of choices available to users narrows and the consequences of a registration denial or cancellation becomes greater.

An Operational Model of IPM

To attempt to define the legal system's role in promoting IPM, it is necessary to have a simple but accurate working model of the concept. There are many definitions of IPM, each stressing the aspect that best serves the interests of the formulator, but all the definitions share at least three common underlying assumptions, which have been well stated by Flint and van den Bosch:

> 1) A conception of the managed resource as a component of a functioning ecosystem -- actions are taken to restore, preserve, or augment checks and balances in the system, not eliminate species. Surveys must be made to evaluate and avoid or diminish disruption of already existing natural controls of both the target pest and other potential pests. 2) An understanding that the presence of an organism or pestiferous capacity does not necessarily constitute a pest problem -- it must be ascertained, before a potentially disruptive control method is employed, that a pest problem actually exists. This requires the implementation of economic injury levels or some other suitable decision-making criterion. 3) An automatic consideration of all possible pest control options before any action is taken -- the integrated pest management strategy utilizes a combination of all suitable techniques in as compatible a manner as possible, i.e., it is important that one control technique not antagonize another (Flint and van den Bosch, 1977)

A number of important consequences follow from these assumptions for the legal system.

First, farmers must be <u>induced</u> to adopt IPM; IPM cannot by imposed upon pest victims. The most obvious reason for this conclusion is that IPM strategies will be crop and location specific, and thus do not constitute a uniform technology that can be fairly imposed upon large classes of the users. The technology-forcing precedents of air and water pollution legislation are not applicable to IPM, for they are premised on the existence of the availability of uniform technologies within broad classes.

Second, in order for an IPM program to be successful, it must be uniformly followed by most, if not all, farmers in the relevant geographical region. Ideally, farmers will voluntarily adopt and police an IPM program but some farmers may hold out. Spoiler "holdouts" impair the success of a program by failing to adopt a necessary practice, thus causing damage to adjacent areas. Besides the spoiler holdout, some farmers may free-ride and thus shift the costs of implementing and managing a program to a group of participating farmers. Free-riders, as economists have shown, may lead to underinvestment in socially desirable activities. Investment is deterred because investors cannot capture the benefits of their investment.

Third, the possibility that there will be spoiler holdouts and free-riders suggests that it may be necessary to impose a program upon an unwilling minority. For better or for worse, our constitutional system is premised on the majority's power to impose its will upon a minority so long as fundamental rights guaranteed by the Constitution are not infringed. Since the 1930s, United States farm policy has been based on grower representative democracy. Crop reduction and marketing programs approved by a majority and imposed upon a minority have long been approved by the courts.

Fourth, EPA cannot be given the primary responsibility for promoting IPM. EPA lacks the requisite understanding of American agriculture and the respect of state pesticide administrators. Because IPM is a locally- or regionally-based concept, the information necessary to gain support for a program must in large part be generated by the state agricultural schools. The difficult question, with respect to EPA is how can IPM considerations be incorporated in EPA's regulatory decision-making, for IPM is a supplement to, not a substitute for safety screening.

Integrated Pest Management is an information system. If the advocates of IPM are correct, it will be adopted by pest victims out of self-interest, for adoption reduces pest costs. One can ask if IPM is in the self-interest of farmers; why can one not assume that the necessary information will be supplied by the market? The answer is that there are insufficient incentives for the production and distribution of the optimum information. Thus there is a case for federal intervention in the form of subsidies to promote the formulation and adoption of IPM programs. To decide how to induce a technology or a process, it is necessary to know why the technology or process is not being produced and used at the rate that is arguably optimum. If the use rate is sub-optimum, adoption of some form of public intervention may be necessary to secure the optimum allocation of pest control resources to IPM. IPM is a sophisticated process that integrates many sources of knowledge. It is not a "nature knows best" or "let nature take its course" theory. It is an attempt to substitute the knowledge of a number of scientific disciplines to achieve least-cost pest control, in place of the current strategy which is, to use an unspeakable metaphor, an attempt to reach a final chemical solution for pests. Thus, IPM depends on the generation and dissemination of a large amount of theoretical and applied information. Not only must this information be produced and broadcast, it must be adopted by pest managers and at the current time dissemination appears partially blocked.

What are the sources of the information blockage? There are many, but the most important are: (1) The state of the art is only in the process of development and there are few programs that can be packaged, as chemical compounds can be, with relatively foolproof, cookbook instructions for effective use; (2) Even if a program could be packaged, it is subject to change from year to year as pest and natural enemy populations may fluctuate. A program may even have to be changed in the years of its adoption due to unforeseen circumstances such as weather; (3) Much important information, which might induce a farmer to implement an IPM program, is not immediately observable and is not sought by farmers. A farmer uses a pesticide to kill a pest. The case for IPM is what the pesticide does not do to adjoining areas, to natural enemies, and the long-term resistance developed by the pest. This information is not likely to be sought by the farmer as long as pests are dying and no crisis arises such as that which occurred in Central American cotton in the 1960s. The costs of the program will have to exceed clearly the benefits before a rational farmer will consider shifting to an IPM strategy, and (4) There is not yet a strong market in IPM information, and there are strong disincentives to the organization of one. Pesticides are made by large companies and marketed through distributors and salesmen. These products are engineered to be as comprehensive as possible to provide the biggest bang for the buck. Heavy reliance is placed on manufacturer representations. A manufacturer has no incentive or little incentive to recommend a program that uses less pesticide nor, it has been charged, an incentive to manufacture selective insecticides that kill a limited range of pests and not their natural enemies.

EPA Consideration of IPM

The Environmental Protection Agency can consider IPM indirectly as an alternative to the use of a chemical registration. And it is the hope of some that the regulation of pesticides will result in a net reduction in the number of pesticides available and the amounts applied each year, thus forcing pest victims to switch to IPM as the only feasible substitute.

EPA's power to consider IPM indirectly comes from its mandate to subject each compound and use to a benefit-risk analysis. During the debates over FEPCA some members of the Senate Commerce Committee demonstrated an interest in promoting the use of IPM. One version of the Commerce Committee's legislation required the EPA to consider "the availability of alternative means of pest control" in deciding if a pesticide posed an unreasonable risk. This language was opposed by the Senate Agriculture Committee on the ground that it gave undue significance to a single factor in the comprehensive risk-benefit analysis. This express authority to consider nonchemical alternatives was washed out in an early Commerce-Agriculture compromise, and the debate turned to the phrasing of the environmental risk standard. The issue of direct consideration of alternatives arose again, but was weakened by a move on the part of the House Agricultural Committee to quash a draft providing that lack of essentiality could be a basis for denying a registration. The Department of Agriculture was developing the policy of registering hazardous pesticides only for essential uses, and the Senate Committee on Agriculture was concerned that a registration could be denied simply because an equally safe alternative was available.

At this point, the Senate Commerce Committee became concerned that an express prohibition against considering essentiality would preclude any consideration of the existence of safer alternatives. Essentiality as defined by the House and Senate Agriculture Committees had taken on a broader meaning than the USDA's definition, and as a result, the definition of the concept had become unclear. In the end, the Senate Commerce Committee gave up trying to make essentiality a criterion for registration or to require an express consideration of alternatives in return for the following assurance:

The language suggested above does not include the phrase "including the availability of alternative

means of pest control" because the balancing of
benefit against risk is supposed to take every
relevant factor that the Administrator can con-
ceive of into account. The question he must decide is "Is it better for man and the environment
to register this pesticide, or is it better that
this pesticide be banned?" He must consider hazards
to farmworkers, hazards to birds and animals and
children yet unborn. He must consider the need
for food and clothing and forest products, forest
and grassland cover to keep the rain where it
falls, prevent floods, provide clear water. He
must consider aesthetic values, the beauty and
inspiration of nature, the comfort and health of
man. All these factors he must consider, giving
each its due. No one should be given undue consideration, no one should be singled out for special
mention, no one should be considered a "vital"
criterion (EPA, 1973).

The enacted version of FEPCA contained the House's insistence
that lack of essentiality is not a criterion for denying
registration, but added the phrase, "where two pesticides
meet the requirements of this paragraph, one should not be
registered in preference to the other" to mollify the Senate
Commerce Committee.

One way to promote IPM would be to place the burden on
the registrant to deny availability of an IPM strategy or to
show that the chemical is needed to implement an IPM program.
However, the legislative history makes it clear that a non-
chemical alternative may be considered in the course of a
benefit-risk analysis, but that the discretion whether to
consider them remains with the EPA. In practice, EPA has
shown a substantial interest in IPM and has cited the availability of IPM as a basis for cancelling a registration. EPA
relied on crop "scouting" in the heptachlor-chlordane cancellation, and in Environmental Defense Fund, Inc. v. Environmental Protection Agency, the court approved the consideration
and reliance upon this alternative:

The Administrator found, with record support, that
no macro-economic impact will occur as a result of
suspending those pesticides. He also found that
crop surveillance or "scouting" for infestations
during the early weeks of plant growth, together
with application of post-emergence baits or sprays
where necessary, provide an effective alternative
to the more indiscriminate prophylactic use of

chlordane and heptachlor. Velsicol urges that this
approach is not as effective as the persistent protection provided by chlordane. Especially in the
absence of proof of a serious threat to the nation's
corn, there is no requirement that a pesticide can
be suspended only if alternatives to its use are
absolutely equivalent in effectiveness. The Administrator reasonably took into account that a transition period would be necessary to implement post-
emergent techniques of control and concluded that
the challenged pesticides could continue in use for
corn protection until August 1, 1976. This evaluation of alternatives and the time required to
implement them is supported by substantial evidence,
and we find no basis to disturb the Administrator's
balancing of costs and benefits (D.C. Cir., 1976b).

Pending EPA regulatory decisions indicate that the need for
a pesticide in an IPM program may also be an important factor in deciding whether to cancel a registration. If such
decisions become EPA policy, a registrant will, in effect,
have to assume the burden of showing either that an IPM program is not available, or that the registration for the con-
tested use is compatible with IPM.

Toward the Institutionalized
Adoption of IPM

To induce the widespread adoption of IPM, the law should
attempt to achieve two results. First, the development of a
pest management industry based on IPM should be encouraged.
Second, the law should allow the creation of IPM districts,
which permit a majority of farmers or pest victims in an
area to implement a comprehensive IPM program (Dunning, 1972).

Licensing and Liability*

The role of the law in the creation of a pest management
industry is to create incentives (subject to appropriate
constraints) to encourage entry to the field, or at least the
removal of entry barriers. Since every new industry potentially imposes costs to the users of its products and services, as well as to third parties and future generations,
society has an interest in minimizing external costs associated with the activity. This interest can be represented by

---

*This section is based on the author's contribution to NAS,
1975.

the standards of liability to which the industry is held.
However, consideration should also be given to regulatory
mechanisms, which assess in advance possible environmental
and other side effects of integrated control.

The law imposes liability for losses suffered as a result
of an activity on two grounds: fault and strict liability or
nonfault. Liability based on fault is imposed if an actor
intentionally causes harm or fails to exercise reasonable
care toward a person to whom a duty is owed, e.g., is negligent. Strict liability is imposed on several grounds. It
was originally imposed by common law for the maintenance of
an ultrahazardous activity. An ultrahazardous activity is
generally one which is abnormal for the particular locale
and one which a court finds likely to result in harm "from
that which makes the activity ultrahazardous, although the
utmost care is to prevent the harm." Increasingly, liability
has been imposed on manufacturers for products with defective
conditions that are unreasonably dangerous to the consumer.
In addition, strict liability has been imposed by the courts
under the sales concept of implied warranty. Warranties may
be imposed either under the Uniform Commercial Code, in force
in all states except Louisiana, or by the courts. It is
unlikely that the UCC would apply to many activities of pest
management consultants because it pertains only to goods and
not services. However, the line between transactions involving the sale of goods and the sale of knowledge is not
clear and has been characterized as "merely a verbal formula
in which results are expressed" for a variety of factors,
such as the need to encourage the activity. However, warranties may be imposed on nonsale of goods transactions by the
courts, for it is generally agreed that the UCC is not intended to stifle the growth of implied warranties of mer-
chantability, and, thus, the courts are free to draw analogies
from the UCC to transactions not covered by it. Under both
tort and sales theories, an actor is liable regardless of
fault, but the defenses available to the defendant differ
under each theory. If liability is imposed on tort grounds,
the manufaturer cannot avoid liability through the use of
disclaimers, but can avoid liability if the user assumes the
risk. If, however, liability is imposed on warranty theories,
liability may be avoided by properly drafted disclaimers.

In three states, Louisiana, Oklahoma and Oregon, liability for damage caused by spraying crops on adjoining
land has been classified as ultrahazardous. In other juris-
dictions liability for injuries resulting from pesticide
use is imposed only if negligence is shown. Pesticide manufacturers have also been held liable for negligence when an

insecticide applied according to instructions caused damage
to a crop, as well as to the target pest species, on the
grounds that the pesticide was misbranded. The failure to
warn about the effect on nontarget crops was held negligent,
and the court held that the pesticide was misbranded.

The utility of a pest management consultant industry and
the likelihood that both consultant and grower will be
equally able to assess the low-probability risks of program
failure, suggest that there is no reason to hold the industry
to a standard of strict liability. In recent years, spreading
loss on an economic base as wide as possible has been advanced
as a basis for strict liability. Some courts have held
defendants liable, on the theory that costs of inevitable
defects should be spread among all users of the product. Loss
spreading is, in effect, a kind of forced insurance for all
consumers. This rationale also seems inapplicable and, in
fact, weighs against holding pest management advisors to a
standard of strict liability.

An industry that performs services, can be indirectly
encouraged by laws that accord it recognition and induce
public confidence by establishing qualifications for the
provision of its services. Such recognition is often accorded
by licensing the activity. The basic legal justification for
licensing is the desirability of establishing a minimum
level of services affecting the public by prescribing standards of practice. Another equally important reason for
licensing is industry desire to increase its status. A less
meritorious legal justification for licensing is control of
entry to the field. Pest control advisors are now licensed
in California, and the principal licensing and license revo-
cation standards are limited to requirements related to personnel competence and to mandatory disclosure to the public
of risks the service entails. The legislation is consistent
with the most restrictive police power justifications for
occupational licensing.

The question will arise whether pest management should
be considered a profession. All licensed activities are
sometimes loosely referred to as professions, but a more
restricted definition of profession is traditionally used by
sociologists and, to some extent, by the law. A profession
is generally defined in terms of association with a body of
theoretical knowledge and a service orientation that is free
from the constraints of the client or the state to define
an acceptable work product. State regulation generally dele-
gates authority to the professional organization to regulate
entry and to establish the standards of practice. The legal

significance of classifying an activity as a profession under this standard is that the courts are more likely to accept internal professional practices as standards defining conduct subject to liability. Members of nonprofessional occupations are more likely to be judged by standards of care external to the occupations. A professional offers a skill and the standard by which this skill is judged for the purposes of imposing financial liability for losses suffered by the users is "the general average of professionally acceptable conduct." The technical significance of classifying an activity as a profession under the restricted definition, is that breach of duty must be established by expert testimony as in the case of medical malpractice. The necessity of using experts to determine if conduct should be subject to liability does not, of course, make the activity a profession. On balance, there seems to be no compelling reason to classify pest management as a profession for purposes of liability determination. It has been argued that the medical profession, for example, should be held to lower standards of liability to encourage medical practice, but there seems to be no reason to hold pest management advisors to standards lower than those applied to any other class of services offered to the public.

The success of a management program depends on uniform participation by all growers within a uniform area. Some form of collective action will be required to organize growers and to compel participation. Districts are superior to other existing techniques because legislative policy is clearly declared, and all members participating in the district have notice of the extent to which growing practices will be curtailed. This contrasts with the more limited powers available to state entomologists to declare that a plant or thing is a nuisance likely to cause imminent danger to the agriculture of the state and to abate it. Such determinations may unfairly surprise farmers and are vulnerable to court challenges based on due process grounds if a prior hearing is not held. Oil and gas fields are subject to compulsory unitization in many states. The laws generally provide that a field may be managed for conservation purposes after a specified percentage of working interest holders approve a unit plan. The objectives of unitization and IPM are similar, and the oil and gas precedents provide a model for fair and socially efficient collective action for pest management (Williams and Meyers, 1977).

A major risk of any district-wide IPM program is that the geographical distribution of costs and benefits may not be uniform. Individual farmers have a constitutional right to be treated fairly and not to have their property taken without due process of law. A farmer involuntarily included in

an IPM district may still object to the fairness of the plan
as it affects him. It is, of course, impossible to assure
that the distribution of costs and benefits will be completely
uniform throughout the district, but an attempt must be made
to do this. Cost and benefit distribution problems can be
best solved by a statute which sets out general substantive
standards for the formation of a district and gives affected
farmers procedural rights to raise fairness objections in advance of the formation of a district. In many cases it will
be possible to modify the plan to meet objections. The issues
raised above have been at the heart of oil and gas unitization
controversies so the standards and procedures developed by
legislatures, courts and administrative agencies provide a
useful precedent for the implementation of IPM.

References

Aidala, J.V. 1978. Regulating carcinogens: the case of
    pesticides. Dept. of Sociology, Harvard University,
    Cambridge, Mass. Manuscript.

Code of Federal Register. 1978. Vol. 40 § 162.

D.C. Cir. 1971. Environmental Defense Fund, Inc. v. Ruckels-
    haus. U.S. Circuit Court of Appeals for the District of
    Columbia. 439 F.2d 584.

D.C. Cir. 1976a. Environmental Defense Fund, Inc. v.
    Environmental Protection Agency. U.S. Circuit Court of
    Appeals for the District of Columbia. 548 F.2d 998.

D.C. Cir. 1976b. U.S. Circuit Court of Appeals for the
    District of Columbia. 548 F.2d 998.

Dunning, H. 1972. Pests, poisons, and the living law: the
    control of pesticides in California's Central Valley.
    Ecol. Law Q. 2:633.

EPA. 1973. United States Environmental Protection Agency,
    Legal Compilation: Pesticides IV: 2026.

Flint, M.L. and R. van den Bosch. 1977. A Source Book on
    Integrated Pest Management. International Center for
    Integrated and Biological Control of the University of
    California.

Kitch, E. 1978. The political economy of innovation in
    drugs and the proposed Drug Regulation Reform Act of
    1978. The Law School Record 24:18 (The University of

Chicago Law School, Winter, 1978) to be published in Proc., The International Supply of Medicines, Sponsored by the American Enterprise Institute, Washington, D.C., p. 15.

NAS. 1975. Pest Control: An Assessment of Present and Alternative Technologies. Vol. III. Cotton Pest Control. National Academy of Sciences, Washington, D.C. 139 pp.

Public Law 6-152. 1910. The Insecticide Act. 36 Stat. 331 (April 26).

U.S.C. 1947. United States Code. Vol. 7 § 136–136y.

U.S.C. 1978a. United States Code. Vol. 7 § 135 et seq. as amended.

U.S.C. 1978b. United States Code. Vol. 7 § 136(c)(5)(D).

Williams, H. and C. Meyers. 1977. Oil and Gas Law. Vol. 6. Matthew Bender, New York.

Zwerdling, D. 1977. The pesticide treadmill. Environ. J. Sept.:5.

Index